Local Projects for 'A' Level Geography

Originally published in 1980, the aim of this book was to help the A-level student make the most of the opportunity to undertake a local project. The general increase of student-activity work in schools was making the project option an increasingly popular part of A-level studies in geography, but there are many problems facing the student in this, the first independent piece of work that he or she is likely to have done.

The most conspicuous difficulty is the time needed to collect data for analysis, but there are many others ranging from the fundamental question of the choice of topic to such subsequent matters as techniques of hypothesis testing and methods of presentation. The author examines all the questions that the student should ask in the course of this type of work, and he does so very much from the student's point of view. The first half of the book deals with selecting the project, setting it up properly and carrying it through to analysis and presentation. The second half consists of a wide range of example projects which illustrate what can be achieved and the problems that arise. They are presented not as models to be copied but as illustrative examples of the general principles and problems discussed earlier in the book.

The range of opportunities available for project work will vary widely from school to school, and no book could provide comprehensive coverage. The author has therefore chosen his material carefully in order to throw light on principles and strategies. The examples are of work actually done by some of the several hundred candidates supervised by the author and they are impressive evidence of the variety of projects that can be undertaken from one home base.

This book has been prepared with the student's point of view very much in mind. It should serve as a valuable source of ideas and guidelines for all those who choose to undertake a project at A-level and it will be useful background reading for students doing similar work in institutions of higher education.

Local Projects for 'A' Level Geography

Alan Pilbeam

Routledge
Taylor & Francis Group

First published in 1980
by George Allen & Unwin

This edition first published in 2018 by Routledge
2 Park Square, Milton Park, Abingdon, Oxon OX14 4RN

and by Routledge
605 Third Avenue, New York, NY 10017

Routledge is an imprint of the Taylor & Francis Group, an informa business

Publisher's Note

The publisher has gone to great lengths to ensure the quality of this reprint but points out that some imperfections in the original copies may be apparent.

Disclaimer

The publisher has made every effort to trace copyright holders and welcomes correspondence from those they have been unable to contact.

A Library of Congress record exists under ISBN: 0049100688

ISBN: 978-0-8153-6279-1 (hbk)
ISBN: 978-1-351-11131-7 (ebk)

Local projects in A-level geography

ALAN PILBEAM

Sir Thomas Rich's School, Gloucester

London
GEORGE ALLEN & UNWIN
Boston Sydney

First published in 1980

GEORGE ALLEN & UNWIN LTD
40 Museum Street, London WC1A 1LU

British Library Cataloguing in Publication Data

Pilbeam, Alan
Local projects in A-Level geography.
1. Great Britain – Description and Travel – 1971-
2. Geography – Methodology
I. Title
914.1 DA632 79-42880

ISBN 0-04-910068-8

Typeset in 11 on 12 point Times by Burgess & Son (Abingdon) Ltd, Abingdon

To the memory of
WILLIAM DELVES

Preface

Candidates in A-level geography normally have the opportunity to submit a local project for examination. Such projects are recommended because they give opportunities for study in depth of particular topics, when the thresholds of textbook work may be crossed. As they do not have the time or topic constraints of examination papers, they tend to emphasise the important qualities of persistence and originality, and they may give sixth-formers their first taste of research. J. S. Bruner writes of learning in general, 'Mastery of the fundamental ideas of a field involves not only the grasping of general principles, but also the development of an attitude toward learning and inquiry, toward guessing and hunches, toward the possibility of solving problems on one's own'. Local projects encourage this development within geography.

Qualities of motivation, independence and creativity are looked for in higher education, and comments on a sixth-former's strength in these are often required in references. All too often, little opportunity is given for these to be assessed at school. A project makes this possible and in addition provides a point of discussion at an interview.

Local project grades at A-level tend to be higher than those for the alternative papers, though perhaps this is not surprising in view of the work involved.

There are, however, difficulties with project work. The most conspicuous is the time required to collect the data for analysis. But there are many others, both theoretical and practical, and this small book has been written to provide help at each stage in the development of a project. It is based on the difficulties encountered by several hundred candidates over many years and it attempts to face these issues from a sixth-former's point of view. Thus problems that loom large in the early stages of a project, such as how to select a topic and how to generate hypotheses, are tackled. The book does not attempt to spoon feed nor to reduce the scope for individuality, but rather to offer advice and encouragement when this is most needed, and to increase the degree of personal satisfaction that may be gained from these small-scale research projects. A number of statistical techniques are discussed as aids to hypothesis testing, and here the aim is to build confidence in handling them rather than to explain them. The book is not a substitute for a good statistics textbook.

Each school will have a unique range of opportunities for project work depending on locality and staff expertise, and the balance of topics selected at a large inner-city school is likely to differ considerably from that of a school in a remote rural area. The experiences considered here from a Gloucester school will, therefore, differ in some respects from all others, and this is inevitable. But the strategies of project work reviewed here are applicable to all sixth-form situations. These are dealt with in Chapters 1 to 6. Then follows a selection of seven example studies, which have been chosen because they illustrate the appropriate scale and depth of A-level projects, and because they demonstrate the application of the principles discussed in the earlier chapters. They show how real sixth-formers have wrestled with real problems, and often comment on why a student has chosen to follow one way rather than another when confronted with difficulties. It is hoped that a sense of the genuine spirit of inquiry can be recognised in these studies. Although example material could have been drawn from many localities, there would have been only superficial merit in doing so. The primary purpose of the book is to explain the methods of project work, not to discuss the results – so the specific area of study is irrelevant. However, the fact that most material comes from one locality illustrates the variety that is possible from one base.

To assist in following the text, OS grid references have been given in several studies. They refer to sheets 162 and 163 of the 1:50 000 series.

I wish to thank my colleagues at Sir Thomas Rich's for their tolerance of many, 'local project seasons' and in particular those former sixth-formers whose work has been used as examples: Miss J. D. Smith, S. D. Allen, P. W. Bright, A. M. Davies, R. E. Day, S. R. Kear, P. P. Montanaro, A. C. Poulton, S. J. Preston and I. P. Simmons. I am also most grateful to Mr J. N. Bates, Mr D. B. Hall, Dr B. J. Knapp, Mr R. F. J. Jones and Mr J. A. Williamson, for their criticism of the script and to Mrs P. Cossins and Mrs B. Irvine for typing it.

ALAN PILBEAM

Contents

1 Some strategies used in selecting a topic

The task of selecting a topic for investigation as a local project is for many sixth-formers a daunting one. It may be possible to select the topic from a list of appropriate themes prepared as examples by an examination board or to persuade a member of the school staff to provide such a list, but to do so reduces the scope for initiative and individuality and undermines some of the educational merit of local projects. So, only consider this as a last resort despite its strong attraction.

The topic that is eventually chosen should correspond preferably to your interests and expertise. The following questions are introduced and developed as guidelines to lead to the most suitable topic; they are not exhaustive and there are many past project titles which would not have been selected using these guidelines alone, but they may prove helpful for students without a title in mind. Of course, if a study theme has already caught your interest, it should be fostered as suggested in Chapters 2 and 3 and this chapter may then be skipped through.

(1) Which section of the geography syllabus has most appeal to you?

In most schools, major areas of the syllabus are left to the upper sixth year. So the experience of work in the lower sixth is necessarily of a restricted range, but this may still be sufficient to indicate whether, for example, a theme from urban geography or from geomorphology has most attraction. Consider the content of the whole A-level course in this selection process by reading carefully through the syllabus, and remember that a theme selected from a basic part of the course will give a deeper insight into this foundation area than will a fascinating but peripheral study. It may be worthwhile to probe your interests further to discover precisely why this particular area of the course has appeal, because the reasons given may be important influences on the type of study eventually selected.

Having identified a branch of geography as the likely topic area, read through your notes or the textbook, searching for generalisations or ideas that could be examined and tested on a local basis. In fact, a small notebook to record such ideas for further investigation is a useful addition to normal subject notes and could be used profitably from the beginning of the sixth-form course.

Suppose, for example, the area of interest is urban geography and in particular the central business districts of cities. Reading through the textbook *Urban geography* by J. H. Johnson (1972, Oxford: Pergamon Press), the following statement occurs in the section on the city centre, 'branches of banks appear to favour a corner site for reasons which are not altogether clear, although local accessibility to neighbouring businesses is presumably important'. Immediately questions spring to mind: Are banks sited in this way in the local area? Have they always been so? Which businesses are nearest the banks? Do the locations of the banks change with changes in the centre of business activity? Such questions would then form the basis of a local investigation.

Possibly one may move directly from the initial topic selection question to one that more clearly sets out a line of inquiry.

(2) Has any item in the coursework suggested a theory that might be applied or a hypothesis that might be tested in the local area?

This is often a fruitful approach, and Study 5, which applies Christaller's central place theory to Beauce, has been chosen to illustrate this type of project.

Each branch of geography has its theories, which may be examined in a local context. In geomorphology the various theories of slope development reviewed by R. J. Small (1977) *The study of landforms* and B. W. Sparks (1972) *Geomorphology,* may lead to a programme of searching for evidence with which to test the validity of the theories for the local area. For example, a project to measure the soil moisture content and infiltration capacity at intervals down a slope and at different seasons of the year would shed light on the level at which surface runoff begins. The theories of Fenneman and Baulig require changes in slope forms at this level. Measurements of soil depth and of slope angles along profiles at the end of spurs and at the heads of valleys would give data to check Gilbert's theory that convex slopes are explained by soil creep. More soil should pass a given point on a valley head profile, where slopes are converging, than at a similar point on a spur, where slopes diverge, so the slope angle should be steeper at the valley head than on the spur, if Gilbert is correct. Strahler's law that, 'within an area of essentially uniform lithology, soils, vegetation, climate and stage of development, maximum slope angles tend to be normally distributed with low dispersion about a mean value...' could be tested by measuring the maximum angles of slope in a local area on different rock types. By plotting on histograms for given rock types the frequencies of slopes at the various maximum slope angles, the statistical distribution of slopes can be checked, and the effects of different rock types can be ascertained by comparing the slope histograms for these rocks.

Von Thünen's theoretical construction of land use patterns around a market could be repeated either with present-day data to show the effects of distance from the farm or market on the type and intensity of land use, or with data from the early nineteenth century gathered from enclosure maps, estate maps and tithe maps, and from the county reports on agriculture. These documents will be available at the county records office, and if estates are selected at various distances from the market town some indication may be gained of whether or not the patterns of farming in Britain followed those noted by Von Thünen in Prussia, and of which variables distorted the zonal layout.

Hoyt's work on the sector-shaped distribution pattern of the most expensive residential property of American cities could be tested with British cities using past and present property values obtained from newspapers and estate agents' records, and by inspecting the size and fabric of the houses.

Every branch of geography has its own theories and models. Some are well established and associated with famous names, others are more tentative, but they all offer scope for testing in a local area. At this stage the method of testing may not be clear and a later chapter tackles this problem. The purpose of this second question has merely been to open up a new range of possible project topics.

(3) Have you access to a special source of information which could provide data on spatial distributions?

Some students, for example, will have access to the records of a family business such as a printing works, or a shop, or a small factory which will include details of sales and purchases. From this information the market area of the business may be delimited by mapping the distribution of sales, and its optimal location may be calculated using weighted co-ordinates. The project may then be developed to consider the constraints that prevent the business from being optimally located – if this is the case.

Conclusions from such projects may have some value to the business, as in the example of a study concerned with the distribution of ice cream sales. Here the numbers of ice creams sold by a van at different sites on its round were recorded over a period of time and sales were related to such variables as weather conditions, child population density, type of housing, time of day and the movements of vans selling other brands of ice cream. The findings were then used to plan a route for maximum sales given the unpredictability of the weather.

A different type of parental information, which was used effectively in a project, was the collection of notes and papers of a planning officer concerned with the design of an urban shopping centre. This provided the details for a comparison of the predicted development of retail activity with the observed activity some five years later. The study then concentrated on the differences between the two and sought explanations for them.

A firm of agricultural contractors specialising in the direct drilling of crops provided the data for another project that mapped the spread of this modern farming technique. The maps were then examined for clues to factors that might explain the technique's diffusion.

The analysis of the environmental factors that correlate with the distribution of juvenile crime has been another project where access to the crime data, though public, was more easily obtained through a parent whose profession was directly involved.

The information handled in these studies will not be available to all sixth-formers, but where similar data sources exist they will be worth considering as important constraints on topic selection.

N.B. Discretion will be needed in handling confidential information and it is probably inadvisable to quote exact figures of sales and purchases, etc. Discretion is particularly important if the study touches on controversial issues such as those of interest to local politicians. It may be worth remarking that, with these issues, a geographical analysis of the facts may not be the *only* appropriate way of tackling the problem.

(4) Have you a hobby, leisure activity, or future career that suggests a problem for geographical analysis?

Occasionally, the leisure interests of a sixth-former lead to expertise, which can be used in very original local studies. The project described in Study 3, which measured the audibility of church bells, illustrates the potential of leisure interests, as does the project on the distribution of rookeries referred to in Chapter 2. The range of projects which might be grouped under the heading of leisure interests is very wide; some of these topics are most successful, others are almost invariably weak. Interests in sport, or local politics, or in railways and canals rarely lead to good studies, probably because the exact data required for spatial analysis are difficult or impossible to obtain, and also because, in the case of railways and canals, so much has been written on their histories that the temptation to rely on second-hand information is too great for many students.

One should of course guard against trivial themes and against topics that are too large. An analysis of the distribution of rock-climbing sites in Britain, for example, is too large a topic for most students.

Career interests may also lead to study topics. A prospective architect has examined the effect of local building materials on rural architectural styles before the railway age brought in cheaper materials from outside the area. A belt transect was made across geological boundaries and such variables as the height and plan of houses, the angle of pitch of roofs, the form of windows and the degree of ornamentation were mapped. A future civil engineer has studied the locations and reasons for slumping along the embankments of the M5 motorway as it crosses the lias clay outcrop, and a candidate for the ministry has looked at the changes in the catchment area of the local Methodist churches, since their foundations in days of travel on foot or by horse.

These studies command a high level of motivation and often entail considerable physical effort. A sixth-former, with a career interest in dairy farming, cycled to eighty farms scattered throughout the county in order to find out why these farmers retained herds of Channel Island breeds in a predominantly Friesian area and in economic conditions that so clearly favour the Friesian breed with its higher milk yields and larger beef calves. And another, with an enthusiasm for archaeology, visited all the thirty-one Iron Age camps on the Cotswold escarpment to check such site factors as area of visibility, access routes, water supply and defensive quality.

(5) Are there local problems which are amenable to a geographical investigation?

Local problems may be local expressions of national problems or problems peculiar to the local area. Examples of such topics are: the distribution pattern of sales of council houses; the changes in connectivity resulting from a reduction in rural public transport services; the delimitation of school catchment areas for secondary education reorganisation; the mapping of atmospheric pollution from a factory chimney; the spread of Dutch elm disease; the design of a more efficient dust-cart collection service; the optimal siting of adventure playgrounds, or clinics, or libraries; and the measurement of accessibility problems caused by the closure of village post offices.

Most editions of the local newspaper will suggest ideas. Here, for example, are headlines from *The Gloucester Citizen* for a June edition:

> *'Extensions dominate plans'.* A list of buildings for which planning consent was sought follows this heading. Details of house improvements and of grants for them may be mapped and their spatial variations within a city examined. These variations may be correlated with other spatial variables such as population density and age structure, social amenities, occupational structure, and the study leads into the delimitation of urban social areas.

'A bid to ban the car from city centre?' This refers to a temporary experiment of closing two city streets to vehicles. Such a plan enables the flow of pedestrians to be monitored both before and after closure. The changes in the flow will show the effects of vehicles on pedestrian movement and will also disclose those shop sites that benefit and those that lose by the greater freedom of pedestrian movement.

Under the heading *'Gloucester Day by Day'*, notice is given of the proposal by the local Association of Family Life to prepare an index of facilities available for children of all ages in Gloucestershire so as to provide a pointer to the places where more efforts should be made. The details of the proposal read just like a local project outline.

'Accident blackspot calls for action'. The geographical factors that are associated with 'blackspots' may be identified with a view to making suggestions for reduction in accident risk at those sites.

The same edition refers to a new road proposal, local footpaths, the effect of the weather on local fetes, the desertion of a mediaeval village, and other items, including house prices, traders' advertisements, job vacancies, which could become project topics. Another day's edition would give another set of headings and another set of ideas for topics.

Considerable interest from outside bodies can be created by these themes and the results of such investigations should be communicated to interested parties particularly when help has been given in providing data.

(6) Can you apply skills acquired in another area of school work to a geographical study?

Frequently a sixth-former's major interest will be in a subject other than geography, and when this occurs the boundary zone between the two subjects may be explored for project ideas. A sixth-former may have expertise in chemistry, physics, or languages, for example, and may be able to use this expertise to investigate a geographical problem.

The streams draining the Cotswolds carry away dissolved calcium carbonate. Various environmental factors influence the rate of limestone solution, and in particular the type of vegetation, the depth of soil, the thickness of limestone, the volume of rain and snow, and the temperature, may be expected to affect solution rates. But one is not able to demonstrate that snowfall increases solution rates or that grassland in winter is associated with higher solution rates than woodland, unless the calcium carbonate content of the streams can be measured. Here the art of the chemist is required in order to titrate the streamwater samples against a 0·01 M solution of the reagent EDTA. An indicator is used to show the end of the reaction and the simple formula:

$$\frac{\text{volume of titre}}{\text{volume of aliquot}} \times \text{molarity of EDTA} \times 10^5$$

gives the calcium carbonate content of the water in parts per million.

Another example of the use of the chemist's skills has been in the mapping of a deserted village. As organic debris accumulated in the village, so the phosphate content of the site was built up. Soil samples were systematically collected from the area and were tested for their phosphate content by the standard colourmetric test using Morgan's solution, ammonium molybdate and stannous oxalate. Variations in phosphate content are shown by different shades of blue, and by mapping these variations the form of the village was revealed.

A botanist has examined the distribution of tree species around the boundaries of woodland to determine which of these woods are ancient and which are recent plantations, and much work has been done by botanists recording the woody species of hedgerows to apply Hooper's theory that hedgerows may be dated by the number of species in thirty yard stretches. In this way, early enclosures have been identified in areas of former open fields and checked with documentary evidence.

Some most interesting projects have been developed from this question, and even studies which ultimately prove inconclusive, such as a linguist's attempt to construct a typical landscape of nineteenth-century Normandy from the novels of Maupassant, are well worth pursuing.

Some of the examples illustrate the use of particular skills. Study 6, which attempted to map the spread of monastic estates, used as one of its sources, *Valor Ecclesiasticus.* This is written in mediaeval Latin.

(7) Can you compare research findings in other areas with those in the local area?

A field course some distance away from home may have given a new awareness of a local feature. For example the problems, arising from pit closures in County Durham, eloquently expressed at a public meeting between residents of a former Category D village and a county planner, created an interest in Forest of Dean villages where coal mines have also closed. Similarly, a fascinating lecture on Durham green villages linked with field work there led to an exhaustive search through enclosure maps for green villages in Gloucestershire, and to the testing of hypotheses about the origins of the large number discovered. An exercise applying Linton's classification of the quality of scenery (*Scottish Geographical Magazine* 1968) to the Nantlle Valley in North Wales, stimulated similar work in the more subdued landscapes of the Cotswolds.

Sometimes the process is reversed, as with the application of some local work on the changes in size and shape of the bedload of a river with distance downstream to the Ardeche, whose course was followed by canoe during a summer vacation.

Reports of research contained in magazines and journals frequently give ideas for local projects. The *Geographical Magazine,* which is readily available, is useful in this respect, but we have found the Institute of British Geographer's publication *Area*, and the journals *Progress in Physical Geography* and *Progress in Human Geography* with their brief reports from geographers working at the research frontier most helpful. These are written with the professional geographer in mind and some papers will be difficult to follow, but they are a fertile source of topic ideas.

Old A-level questions can also be adapted for project titles. 'What factors influence the proportion of total precipitation that seeps into the ground?'; 'Why do some towns grow asymmetrically?'; 'Is variation in rock type the main factor influencing vegetation on a local scale?' – are three examples of questions that lend themselves to local investigation.

A careful and thoughtful consideration of all seven questions is recommended, for they are guidelines that have proved to be helpful to many candidates. Then discuss your ideas with the school staff. Their knowledge of the opportunities of the local area will be invaluable, and it may be worth a surreptitious inquiry as to what their own special geographical interests may be!

The final selection of the topic will be influenced to some extent by your preference for working environments – in the library, the laboratory or in the field – and by your preference for dealing with maps, statistics, or people.

2 The definition of the problem

Within the framework of the selected topic, the aim now is to specify a problem for investigation. In most cases the problem will be clear already, but for some students further guidance will be needed.

A problem exists when there is a difference between what one expects to find and what is actually found. In geographical studies this conflict is often between an expected and an observed *distribution pattern* and one is led to ask why this or that difference occurs; or it may be between an expected and an observed *form* as, for example, in Study 1 where the form of a slope did not correspond to the textbook diagrams. Sometimes the conflict is obvious, in other cases it is more subtle and is only recognised after considerable work has been done. It will be noticed later that most of the example studies uncover new problems for investigation in the course of attempting to solve the initial one. In fact, T. S. Kuhn in his important book *The structure of scientific revolutions* (1970, Chicago: Chicago Univ. Press) argues that the progress of science depends on such conflicts and that they arise inevitably as instances occur that cannot be explained by accepted knowledge.

It is worth noticing the significance of expectation. An expected distribution pattern is really the product of the application of one's geographical knowledge to the phenomenon under review. It is, in a sense, a summary of one's existing geographical knowledge and it clearly warrants careful thought. For example, in a study of the intensity of land use on a farm, one factor likely to be highly important is the accessibility of the different fields, which is usually measured as distance from the farmyard. One would expect that land use intensity would decrease with increasing distance from the farmyard, and so some prediction of the effect of distance on such variables as the numbers of days of farm labour (standard man days, SMD) should be made before checking them with reality. The check will identify anomalies and these then become the problems for investigation.

A simple way of identifying anomalies is through the construction of graphs. In the example above, the distance of each field from the farmyard would be plotted on the X axis (the independent variable) and the SMD required in the fields would be plotted on the Y axis (the dependent variable). One would expect to find an inverse correlation with the SMD decreasing as the distance increases and the graph would then appear as in Figure 1.

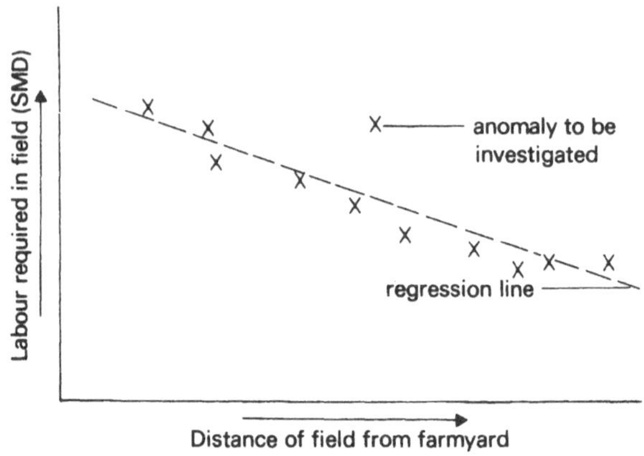

Figure 1 Graph showing relationship between distance from farmyard and labour intensity.

When a regression line (line of best fit) is fitted to the scatter plot, the anomalous fields appear as those dots farthest from the line. Attention can then be focused on these.

It may be necessary to show that there is a problem to be solved, for example, to show that the variations occurring in a distribution pattern are not just a matter of chance. A rather unusual biogeographical study well illustrates this procedure and the chi square (χ^2) test that is involved.

The distribution of rookeries

Owing to the spread of Dutch elm disease, the numbers of nesting trees for rooks in the Severn Valley have been severely reduced and rookeries have been moved. New sites, however, must still be within range of feeding grounds so that the parent birds can make the frequent flights back to feed their young. Before investigating the effect of the distribution of corn land on the new rookery distribution, the study had to show that there were variations in rookery density, and that the pattern was not random but one that hopefully would reflect the land-use pattern. The sample area of study was divided into the different land use belts and the number of rookeries in each belt was recorded. Knowing the total number of rookeries in the study area of 350 km^2 and the proportion of the whole area in each belt, it was possible to calculate the number of rookeries that might be expected in each belt if the distribution was uniform. The results were as follows:

Land use belts	*% of area*	*No. of observed rookeries*	*No. of expected rookeries*
Forest of Dean	8%	1	9 (i.e. $114 \times \frac{8}{100}$)
Severn Valley marl	12%	22	13
Severn Valley alluvium	11%	11	12
Severn Valley clay	28%	38	32
Cotswold scarp face	7%	10	8
Cotswold dip slope	34%	32	40
	Total	114	

The formula for the χ^2 test is:

$$\chi^2 = \Sigma \frac{(O—E)^2}{E}$$

where O is the observed frequency, E is the expected frequency and Σ is the summation sign. Using the above data:

$$\chi^2 = \frac{(1-9)^2}{9} + \frac{(22-13)^2}{13} + \frac{(11-12)^2}{12} + \frac{(38-32)^2}{32} + \frac{(10-8)^2}{8} + \frac{(32-40)^2}{40}$$
$$= 7{\cdot}1 + 6{\cdot}2 + 0{\cdot}1 + 1{\cdot}1 + 0{\cdot}5 + 1.6$$
$$= 16{\cdot}6$$

Using χ^2 tables, follow the row for df = 5 (there are six land use belts so there are 5 degrees of freedom, $n-1$) and in the column 0·01 read the value 15·09. This gives the value required for the null hypothesis (that there are *no* statistically significant variations) to be rejected at the 99% or 0.01 confidence level. The calculated value of χ^2 is greater than this so the null hypothesis is rejected and the investigation can proceed, because it has been shown that there is a statistically significant variation in rookery density between the different land-use belts. Incidentally, the same test was also applied to the numbers of nests and with a similar result.

The wording of the problem

The wording of the problem may now be considered. Normally the problem is stated as a question and as such makes the title of the study. Four comments may be offered as a help in the construction of the question:

(a) It should be *clear and precise.* Ambiguities easily occur and may be avoided by stating briefly the type of answer the question logically demands. If this is the type of answer given by the solution of the problem, the wording of the question may stand; if not, a rephrasing is necessary.
Consider, for example, the suggested problem, 'How successful has recent office building been in Gloucester?' The question arose from an awareness of extensive office building programmes in many towns in the early 1970s and from an observation that some were not immediately occupied. But, as the question stands, it is not geographical and it could be answered in a variety of ways. An architect would reply in terms of the design of offices, an employment officer in terms of the numbers of jobs created and a financier in terms of rent values. The geographer wanted to know if the sites of the office blocks affected the delays in letting, and so a more precise question was constructed, 'Where and why have new offices in Gloucester remained unoccupied?' A colleague, especially a non-geographer, could give a useful second opinion on the clarity of the question.

(b) It should be *manageable,* and preferably *solvable.* A small-scale rigorous piece of research is normally of much greater value than a large study which, because of its ramifications, is diffuse and vague. The question may indicate the reduced scale by referring to a limited area, sample, or restricted time scale. For example, in Study 4 the sub-title mentioned the sample area of study within which settlement siting factors were to be investigated. On the other hand, the study should not be too constricted and advice should be sought if there are some doubts. The example studies have been chosen to illustrate the appropriate scale of A-level projects. Most examination boards require the submission of study outlines before a project is accepted, and comments by the examiners may be made on the suitability of the study and on some of the difficulties likely to be encountered. Confidence in academic work is achieved through success, hence the importance of the manageability of the question. The eminent scientist P. B. Medawar stresses this point when he writes, 'No scientist is admired for failing in the attempt to solve problems that are beyond his competence . . . research is surely the art of the soluble'.

(c) It may be *convergent or divergent.* Convergent questions lead to one answer, divergent questions to many answers. 'Which local farming enterprise is most efficient in its energy budget?' is a convergent question, whereas the question tackled in Study 6, 'What factors may have influenced the spread of monastic estates?' is divergent. Some studies involve both. The psychologist J. P. Guilford has suggested that some people tend to think convergently, others divergently. Convergent questions give scope for precision and analytical work, divergent questions perhaps encourage more imaginative and creative thought. Both convergent and divergent questions may be equally demanding and equally suitable for local projects, but one should be aware of their different ends.

(d) It should be *relevant* to the subject and to the investigator. The first chapter covers personal relevance but some further remarks may be made about the former point. Historical data and geological data are often used in geographical studies, as of course are data from many disciplines, but the data must be applied and not left undigested. The history of a parish or of a railway line is not geography, nor is the stratigraphy of the Jurassic period. By the careful wording of the question some safeguard against irrelevances can be made. Geographical questions normally refer to a distribution pattern or have some other spatial aspects.

Several of the examples contain additional ideas on the definition of the problem. It is often said that the art of research lies in asking the right questions, and this is certainly true with local projects.

3 The formation of hypotheses

It is probable that you have some notion as to what might explain the anomaly or the problem that is to be tackled. Hopefully, some past experience in geographical study will hold clues. But it is just conceivable that you have no idea. So a few comments on sources of ideas may help.

Field observation

With most studies it is essential to go into the field and to observe carefully the problem area. Look at it from different viewpoints, perhaps in different light conditions, or in different weather conditions, or at different times of day. It is often surprising how one's impression of a street, town, or landform can vary with different viewing conditions and how changed conditions suggest new ideas. Some of the most satisfying local projects are those for which an absence of background knowledge has forced the student to observe more closely than would otherwise have been the case. The first study, which is concerned with the origin of some minor landforms, illustrates the value of this approach.

Another example of the importance of close observation is found in the study of the changing functions of a city street, where the modification of building frontages may give clues of former uses. A cursory glance through the crowds of shoppers on a Saturday afternoon will not be sufficient to disclose the alteration of a window or doorway, or the contrast between the structure of the ground floor and the first floor. A methodical, building-by-building, visual inspection is required. Some processes in physical geography are spasmodic, and landforms may change conspicuously over short time periods and then remain comparatively stable for a long time. For example, a river at bankfull discharge occurs rarely, perhaps once or twice a year, and it is then that the channel forms change rapidly, and that observations will produce ideas for later testing. The reasons for the desertion of a mediaeval village site may not seem obvious on an idyllic June afternoon, but one at least is apparent in March with an icy north-easterly wind delaying the growth of corn and grass on the exposed slopes.

Observation in the field will not only suggest ideas to be developed as hypotheses, it will also act as a useful check on the feasibility of those hypotheses, which may have originated, for example, from map work or from statistical work, where a correlation between the distribution patterns of two phenomena has been found. Further remarks on this function will be found in Study 4. So a preliminary leisurely observation of the problem area should be made.

Background reading

Ideas also come from reading about the topic under consideration. Begin with the relevant chapters in the standard A-level or undergraduate textbooks and then systematically follow up the references given in them. Not only is this background reading necessary to master the subject, it is also a primary source of ideas. As an illustration of this use of references we may take the problem of the origin of the large valley meanders of the Cotswold dip-slope rivers, e.g. R. Coln (GR 0217 0317). The relationship between the mean meander wavelength and the bankfull discharge of a river can be calculated, and in the case of the present day Coln wavelength (m) $\simeq$ $11\sqrt{\text{discharge}}$ (cumecs). This relationship can then be applied to give the approximate discharge required to produce meanders of the size shown in the valley contour pattern. The problem, then, is how could such a large river be formed in such a small drainage basin? A problem made more acute by the equally large meanders in adjacent valleys. A search through the literature on river discharge and on meander formation would be needed at this stage. Several geomorphologists have worked on the problem and their findings have been published, sometimes in local journals of limited circulation. This search would yield the following possible explanations: that the river was the overspill of a glacial lake north of the Cotswolds; that the drainage basin was once much more extensive but river capture has beheaded the Coln; that the large river resulted from snow melt above a permafrost layer in periglacial conditions; that the river was produced in conditions of heavier or more intense rainfall; that the present day Coln is reduced in volume because an impermeable layer of rock has been breached and water now infiltrates; that the valley meander forms were determined by factors other than the river. Some of these hypotheses could then be taken up for further investigation.

Some of the references may be to rather obscure papers, but this does not mean they are unobtainable. For books and papers dealing with purely local information and for the journals of local learned societies, you might consult the reference library of the nearest city or the county records office. For reference to articles in geographical journals it may be possible to visit the libraries of a college of education, polytechnic or university. Librarians here are often most helpful, but the reading will have to be completed within the reading room. Failing these approaches it may be possible to borrow the journals from the Geographical Association or from the Royal Geographical Society, if the school has a link with these bodies. Texts may also be obtained in this way, but the simplest and cheapest way is through the interlibrary loan scheme. It is very unlikely that given persistence and time a reference cannot be obtained. You do not have to live in a university city.

Discussion

Discuss the problem with other people, with other members of your group or with anyone who will give a sympathetic hearing. The very act of expressing a problem in conversation helps to clarify one's own thought about it. But be prepared to have pet ideas dismissed in somewhat cavalier fashion. It can be disconcerting to a student working, for example, on the analysis of agricultural land use, who has considered the effects on land use of distance decay, field size and shape, slope angles and aspect, land drainage and so on, to be told by the farmer to forget about these, 'It's the price review that matters'. In other words, to the farmer the biggest single factor influencing land-use decisions is the guaranteed price of the product, which is ultimately determined by world markets and hence often by the harvest conditions in the USA the previous year; hardly a local factor!

Most places have their 'local expert' who may be worth talking to. In a village an inquiry at the post office, rectory, or public house will often put you in touch with this person. In an urban area the channels may be initially more formal, such as through the museum, library or planning office. When dealing with these often busy people, have some questions well prepared before the interview and show that you have already given careful thought to the problem.

Other aids

There is some research evidence from psychology that a visually stimulating environment facilitates hypothesis generation, and also that a change of consciousness, as with sleeping on a problem, may help. Exactly how ideas are created is not known, but it seems that time to mull over the problem at leisure assists in their production. One does not need to have a high IQ to come up with a bright idea, though this may help in dealing with the idea once it has arrived!

When a number of ideas occur to you, some method of checking their coverage may be necessary. Projects evaluating alternative uses for a moorland area or a derelict site, such as a disused railway station or a block of condemned houses, illustrate this point. A variety of possible uses will be considered, but their relative merits depend on the projected time scale for development – one year, ten years, fifty years; and on the spatial scale of evaluation – locally, regionally or nationally. So a simple matrix could be drawn with a three-point time scale on one axis and a three-point spatial scale on the other, and ideas could then be inserted into the relevant boxes (Fig. 2). In this way gaps in the coverage are soon apparent, and further thought may be directed to these areas.

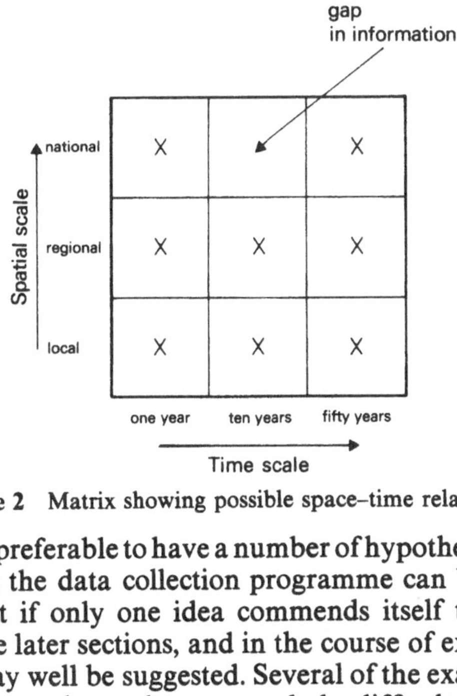

Figure 2 Matrix showing possible space–time relationships.

Clearly it is preferable to have a number of hypotheses before testing begins so that the data collection programme can be planned more efficiently, but if only one idea commends itself to you, test it as outlined in the later sections, and in the course of examining it other hypotheses may well be suggested. Several of the examples show how new ideas arose as the work progressed; the difficulty then is to decide which to leave aside.

Hypothesis generation: a case study

As a summary to this section the whole strategy of hypothesis generation will be reviewed in a case study.

Recently, attention has been drawn to the plight of the English country house and an examination of the distribution pattern of country houses in a sample local area was undertaken in European Architectural Heritage Year. Leaving aside the initial problems of determining the houses to be included in the term, country house, and of the demonstration of their varied distribution pattern, the sequence of stages in the quest of location factors to be considered as hypotheses was as follows.

First, their sites were marked on the 1:50 000 OS map, which was then inspected for common site factors. This search revealed the frequency of water courses flowing through the grounds. It also revealed areas where country houses were sparse – crown land, alluvial land and scarp slopes, for example.

Secondly, each house was visited and the same impression of the probable significance of water was gained. Other points were noticed too. The range of architectural styles indicated different dates of construction or reconstruction. Did the type of site chosen at one period differ from the type of site selected at a later period? The building materials were seen to vary. Was the availability of suitable building material a location factor? One house was constructed on the site of a monastery. Did the dissolution of the monasteries influence the distribution pattern of country houses? Many of the houses visited now had other uses such as hotels, conference centres and schools. The significance of this observation was not initially apparent but it later took its place in hypothesis construction.

Thirdly, the local library was searched for books on country houses. Several were read and a number of issues of *Country Life* and the *County Magazine* were browsed through for information. Some early details including eighteenth century engravings were found in Sir Robert Atkyns' *The ancient and present state of Glostershire* 1712, and the county volumes of the Penguin *Buildings of England* series gave more. The practical comments of a seventeenth century builder, Sir Henry Wotton, were discovered. The site 'must not be subject to any foggy noisomness, from fens or marshes neer adjoining ... Not indigested for want of sun, nor unexercised for want of wind'. He wrote that 'vast and indefinite views' are condemned by good architects. And then he gave a useful 'political hint': 'Do not build too near a great neighbour as it will mean living on Earth as Mercury is in the Heavens – ever in obscurity under brighter beams than his own.'

From these readings a number of new questions arose. What were the minimum distances between country houses? Were Wotton's comments about views typical of his day and therefore of the houses built at that time? Did the change in garden fashion from the seventeenth century formal garden to the eighteenth century landscaped garden influence siting in those periods? Was seclusion a location factor?

Discussion raised more ideas. Did the source of the families' wealth influence location? A change in the relative agricultural prosperity of different parts of the area might be expected to affect the country house distribution pattern. So would developments in industry – iron working, coal mining, woollen manufacture. Discussion also drew attention to the importance of access to the social life of London. Were there more country houses in the direction of London? Were they located along the coach roads?

Then the local newspaper published an article on the dereliction of one of the local houses and so introduced a new dimension. How much does the present pattern owe to past destruction? Were there other houses in the area at earlier periods? Did their location affect their chances of survival? How important was the location of a country house for the possible alternative uses that prevent the demolition orders?

And so the list of ideas was extended. Next the ideas were prepared for statement as hypotheses. For example, that sites selected in the seventeenth century differed from those of the eighteenth with respect to extensive views; that sites were chosen along the roads to London; that country houses were located in areas of agricultural prosperity. Finally, the project systematically tested these hypotheses. The sequence of stages is shown in Figure 3.

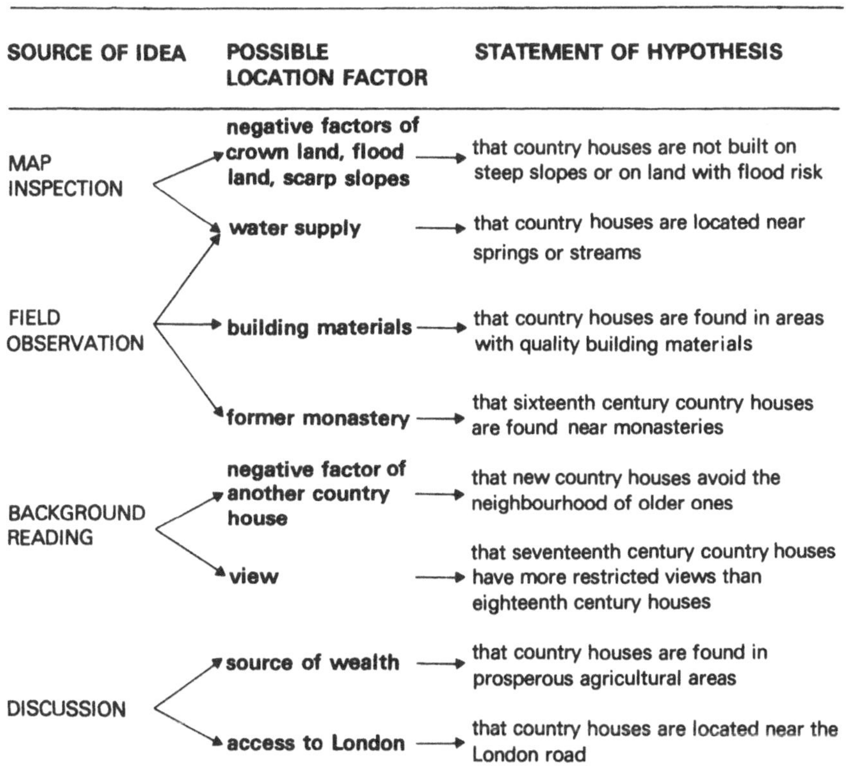

Figure 3 Diagram showing stages in hypothesis formation for a project on country house locations.

During the hypothesis generation stage of the project and for all subsequent stages a notebook should be kept to record the progress of work. Not only will this assist in the writing of the study at a later date, but it also monitors the process of geographical investigation and, if produced at the examiner's interview, will show aspects of the work that the written account may not include.

4 Comments on the collection of data

Once the hypotheses have been constructed, the nature of the data required to test them will be evident. For some projects the ideal data are not readily available, and it may be necessary then to use sub-optimal information; but the precise factual requirements for the testing of each hypothesis should be clearly stated.

Hypothesis constraints

This point is emphasised not only because of its logical priority, but also because it focuses attention on the quality of data which are available. Unless the required information is stated, the inadequacies of the available information may not be fully appreciated and so the validity of the hypothesis testing may be over-estimated. One or two examples will demonstrate how easily this weakness may occur in a project. Studies concerned with variations in the growth of vegetation on land at different altitudes or with different aspects may require temperature readings, but the readings required are likely to be those of the actual environment in which the plants are growing, not those taken in a Stevenson's screen some 1·2 m above the ground surface. Studies investigating the contrasts between social areas in a city may require population data, but the available data published in the census reports are produced on a ward basis and the boundaries of the wards may not correspond to those of the social areas being studied. Here again the data may be deficient, and for more accurate work it will be necessary to collect one's own information or to reconstruct the hypothesis so that it fits the type of data available. Where the rate of erosion along a stretch of coast is being measured and some explanation of spatial variations in that rate is attempted, it is probable that geological data will be needed. The structural details recorded on the one inch geological map are of too small a scale to be of much help with this type of fieldwork problem, the required data includes details of many minor changes in geological character, which can only be measured in the field. So give adequate thought to the exact information that you need to test the hypotheses and set out the requirements as clearly as possible.

Technique constraints

The other important constraint on the type of data to be collected is the technique with which it is proposed to test the hypothesis. Frequently it will be necessary to calculate a coefficient of correlation which describes the relationship between two sets of data. If one of these data sets cannot be measured on an interval scale but can be arranged in rank order, the statistical technique to be used will be one of rank correlation. There is then little point in striving after precise measurements for the second set of data when the rank order alone will suffice. The first Study mentions an exercise where the relationship between rock hardness and maximum slope angle is assessed. Home-made clinometers or pantometers will give accurate slope readings, but rock hardness is more difficult to measure. Rocks, however, may be ranked in their order of hardness by hammering them, and the Spearman rank correlation test may be used. This means that the rank order of maximum slope angles on the different rock types is all that is required, not the precise angles. On the other hand, if accurate measurement of both sets of data is possible then a more powerful correlation test may be used such as the product moment correlation. Each statistical test has its own requirements about the data to be used and much time may be saved when these are checked before the data collection begins. Some reference to these requirements will be made later but the standard statistical texts such as Hammond and McCullagh (1978) should be consulted on this matter.

Sampling constraints

Probably the greatest problem in data collection is concerned with sampling, and no doubt most A-level projects break some (or all!) sampling rules. The purpose of sampling is to save time and energy in data collection and to enable valid conclusions to be reached about the whole population (whether people, pebbles or pubs) from a representative number. Two common questions arise here: how large should the sample be, and how should it be selected? The first question is more difficult to answer because it depends on the problem to be solved, on the range of the data, on the degree of error which is acceptable, and on the methods of data treatment which are envisaged. Several general comments will be made before discussing examples. Where parametric tests are to be used (these tests assume a normal population distribution) such as the product moment correlation, and where one is not sure of the normality of distribution a minimum sample of 30 would be acceptable. For the chi square test the minimum number is 20 and no expected frequency should be less than 5, and for the Spearman rank correlation, at least 7 pairs of readings are required. Note that these are minimum numbers and larger samples collected wisely will give a reduced margin of error. Obviously, it is much easier to collect data for some projects than for others. A study of supermarket locations in a city is likely to involve smaller numbers than one of pebble distribution patterns on a beach, and the sample for the latter would be expected to be larger. It is often helpful to calculate the standard error of a sample, which is a statistical measurement of how confident we can be in drawing conclusions, from the sample, about the whole population. Again, a statistics textbook will explain the procedure.

The statistical tests to be used also determine the method of selecting the sample. Where parametric tests are to be used, the data will be selected randomly. For most tests, in fact, a random sample is to be preferred, but in some cases the simplest and most efficient way of collecting data is by a systematic method. In the rookery study referred to in Chapter 2, information on land use was needed. A random sample of fields could have been taken to give the land use data by covering the area with a grid and then using a random number table to give the figures for co-ordinates to locate these fields (Fig. 4). But the travel involved to visit each field to find out the land use would have been excessive. A simpler approach – the one adopted – was to record land use along a number of parallel lines crossing the area. Ordnance Survey grid lines were used thus giving a regular systematic sample. If randomly selected data had been necessary, the grid lines could have been randomly chosen and points along them randomly selected.

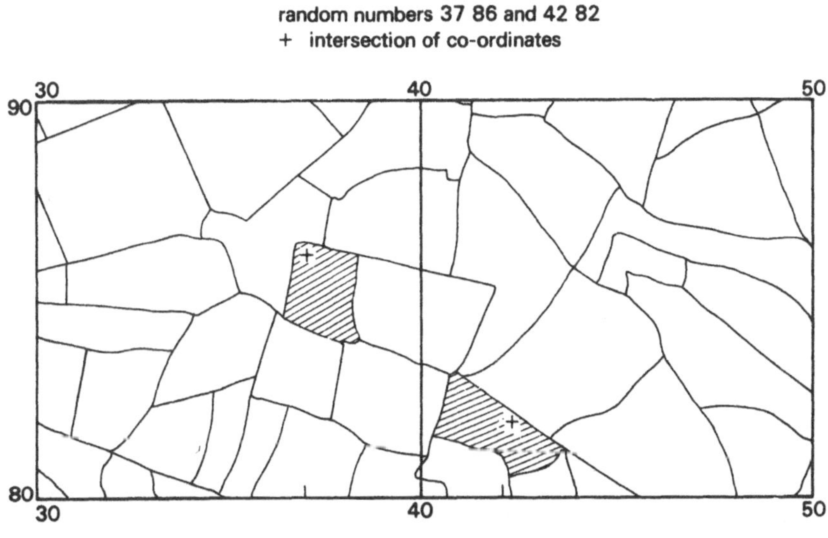

Figure 4 Random sampling of fields.

One of the aims in random sampling is to avoid bias in the collection of data. The colourful or conspicuous or rare item is apt to attract undue attention in data collection and this is avoided in a random sample. Walking across the terraces of a river valley one may notice that the surface deposits vary in the size and material of their pebbles, perhaps indicating that the river had a different competence with each terrace level and that the catchment area of the river had also changed. Before it can be shown that there is a significant difference between the pebbles of each terrace, a random sample must be collected from each level. Picking up thirty or so pebbles will not give a sample that can be *shown* to be random. Larger and more colourful stones may predominate, or if one is conscious of this likely bias the selection will include more small and dull stones than is representative. To be objective a quadrat, such as is used in botanical sampling, should be laid across the pebbly surface and the sample taken by random number co-ordinates. Once this random sample has been collected, the measurement of the pebbles can begin and the variations between deposits ascertained.

A similar need to demonstrate that an unbiased sample has been used may occur when the results of questionnaires are being analysed. Some people are much more approachable than others, and unless a random sampling technique is rigidly adhered to, it is the people with kindly faces whose opinions and information will be recorded and this may not be representative of the population being studied. A random number sequence could be used to indicate which people are to be approached and such procedures might be adopted, for example, when testing whether there are differences between the types of customers using shops located at varying distances from a city centre.

Published data

There are many sources of published data which local projects can draw from. Several of the frequently used sources are listed on page 105 and these are normally available at the local reference library, county records office, or local government offices. With all published data it is important to check the methods used in collection to see if they are appropriate to the project in hand. A number of other sources are used less frequently and may be referred to here in order to stimulate the search for similar sources, which may be helpful to your own particular project.

The *Farm management pocketbook* compiled by John Nix and available from Wye College, Ashford, gives information on farm prices, labour requirements, machinery costs, and many other details needed for planning the work of a farm. These facts can be most helpful to local studies dealing with agricultural topics and with analyses of rural land use patterns. For example, an estimate of the relative efficiencies of farms in different localities can be made using information on acreages of crops, head of livestock, capital equipment and labour provided by the farmer, and converted into monetary terms with the aid of the pocketbook. Then the possible causes of these differences can be tested as hypotheses.

The *enumerators' schedules* for the early census returns may be of help in detailed studies of population change. From 1841, details of the age, birthplace and occupation of all members of the household are recorded, and provided the enumeration districts can be identified, this information can support, for example, a study of the effect of the railways on rural settlements. Some villages had railway stations, others did not. Was there a growth of population in the railway settlements? If so, where did the people come from; and what was their occupation? The schedules give the answers. Many towns such as industrial towns, spas and resorts grew rapidly in the first half of the nineteenth century. The schedules show where these people came from. The county records office will probably house at least one set on microfilm. There is a hundred year restriction on their public availability so only the span of 1841–71 can be consulted at present.

The county records offices hold extensive documentary resources such as estate records, enclosure maps, tithe maps, parish records, sales catalogues and plans of early routeways. Projects which contain an historical theme such as the evolution of a settlement, the siting of early industry, the development of field patterns or of land holdings would be able to draw on these resources. They will be catalogued and the archivists are always prepared to give advice on the availability of early documents.

Although the *Yellow Pages* of telephone directories assist in studies of contemporary economic activity and are frequently consulted for the locations of businesses, earlier directories also have value. Christaller, in fact, used telephone directories as a data source in order to demonstrate central place theory in Southern Germany, and the Post Office may have copies of early directories, which can be put to similar use to measure the centrality of places. A study which has used a 1938 directory tackled the question, 'Does Gloucester benefit or suffer from the nearness of Cheltenham in terms of business activity?' As this directory, which covered Gloucestershire, Herefordshire and Worcestershire, only contained about fifty pages it was relatively easy to extract a list of businesses for Gloucester, Cheltenham and Worcester. In 1938, Worcester had a population approximately the same as that of the other towns and, as it was relatively isolated and independent, it could be regarded as a control. Some types of business occurred in equal numbers in all three towns but for many others marked differences were shown.

Another interesting documentary source is the show catalogue with its addresses of exhibitors. This has been used with the Three Counties Agricultural Show to plot changes in the distribution patterns of livestock breeds and in the catchment area of the Show. These changes reflect those of mobility, farming practice and breed popularity.

Study 6 illustrates the use of another source, *Valor ecclesiasticus,* which gave information on the monastic holdings at the time of the dissolution of the monasteries and the necessary background details for a study of the geographical factors that influenced the spread of monastic estates.

Reflection on these less obvious published data sources will be sufficient to suggest many more which could be consulted in project work.

Instrumental data

Projects in physical geography will often require the collection of data that involves the use of instruments. There is scope here for a technically minded student to design and construct his or her own instruments. The use of thermocouples for measuring the temperature of different surfaces for a micro-climatology project may be taken as an example. Two pieces of copper–constantan wire, a thermos flask of ice, and an potentiometer borrowed from the physics department to measure EMF are needed. With this equipment suitably calibrated, instantaneous readings to a fraction of a degree are possible. Surface temperatures around buildings or in a street may be measured and related to weather conditions, or to the flow of air as traced by large soap bubbles; and temperatures in a hedgerow or woodland ecosystem may be related to vegetative growth and dates of flowering or leaf opening.

Several sources of details on instrument construction are available and we have found *Techniques in physical geography* by Hanwell and Newson (1973) particularly useful, and some of the GA *Teaching Geography Occasional Papers* also contain advice on instrument making. There is plenty of opportunity for ingenuity in measuring the physical environment, but a word of caution may be needed in handling the data collected. Some awareness of the range of experimental error should be shown in the study in the same way as sampling error is recognised in statistical analysis. Calculations to the nth decimal place may not be appropriate with all instrumental readings.

Group work

The collection of data is often a tedious business and the monotony may be eased by sharing the task with other students. This sharing applies particularly well on field courses where, perhaps, one whole day is devoted to a specific problem and a large group collects data, which will be processed later by one person. Another simple illustration of this is when a long journey on a school minibus can be put to use by recording the crops grown in fields at regular intervals along the route. A rota of recording, crop identification, and mileage reading is drawn up and so the work is shared. Such transect details have been used to test whether or not the traditional *pays* of the Paris Basin can be identified from land use patterns, or whether the regional divisions of Belgium are accurately portrayed in the textbooks. The open-field landscapes of the North European Plain are well suited to this method, although the land use bordering the main roads may not be typical of a region and some sampling off the roads may be necessary. Standardisation of recording method is, of course, vital when a number of different people are employed in data collection.

Questionnaires

Data collection by questionnaire raises many problems. Apart from the sampling decision of who should be approached to answer a questionnaire and the difficulties created by a low percentage of returned questionnaires, there is the initial problem of wording the questions. So often the words of a question condition the response. In assessing the advantages and disadvantages to a local community of mineral working, such as limestone quarrying, the first question to a householder, 'Do you suffer in any way from the quarrying and if so, how?' immediately raises the objection side of the issue. Careful thought should be given to the wording of questionnaires. With the plethora of local studies around schools, and especially in the popular field work areas, questionnaires should be used with great discretion and, if there is any doubt about their value to a project, do not use them. They are, however, essential to a few studies and here the aim should be for politely worded, simple, precise and factual questions.

Time constraints

Almost inevitably, it seems, some projects are handicapped by a shortage of data. Sometimes promised information does not materialise or a surrogate has to be used for the ideal data such as rateable value for land value, or pH for soil fertility. Difficulties like these cannot be avoided, but others can. Studies dependent on weather conditions, for example, should be begun at an early date in case a period of drought or unusually persistent rain or severe frost prevents data collection for a time. How many studies of hydrology have coincided with dry weather, and projects on frost incidence with mild winters? Some business enterprises, in addition to farming, have marked seasonal changes and here again data collection should be spread over as much of the year as possible. It should be obvious which projects are likely to require this longer period of recording and, unless there is the time available, these topics should be avoided.

5 Techniques used in hypothesis testing

There are many different ways of testing hypotheses and this chapter reviews the most commonly used techniques available to sixth-formers. The majority are statistical tests but these are simple and easily followed, and the calculations they involve would be within the range of an able first-former. They may however, be tedious, and a pocket calculator would save much time and effort in performing the calculations.

By far the most frequently used method is that of correlation, where one aims to describe the relationship between two sets of variables, the one being the item under investigation and the other, its hypothetical 'cause'. For example, the distribution of vandalism in a city might be correlated with the distribution of derelict buildings; or the number of different species of plant with the ages of the spoil tips on which they are growing; or the grade of the bedload of streams with their maximum discharges.

Graphs

As a preliminary, the two sets of variables should be graphed. It is conventional in the plotting of graphs to place the independent variable – e.g. density of derelict buildings, ages of spoil tips, stream discharges – on the horizontal *X* axis, and the dependent variable on the vertical *Y* axis. There are several reasons why the construction of graphs is so valuable in hypothesis testing:

First, the nature of the relationship between the two variables is shown visually. Where the readings along the *Y* axis increase with those along the *X* axis, a positive correlation is shown, as when rainfall readings increase with altitude. With a negative correlation the *Y* readings decrease with increase in readings along the *X* axis, as when the amount of labour on a farm decreases with increased distance from the farmyard (Fig. 1). But many relationships are not shown by simple straight line trends. For example, in the project measuring the calcium carbonate ($CaCO_3$) content of a stream draining the Cotswold limestone, one might expect that the carbonate content would increase with distance downstream. It does, but the trend line on the graph is curved (Fig. 5). Solution of $CaCO_3$ increases rapidly at first, but tails

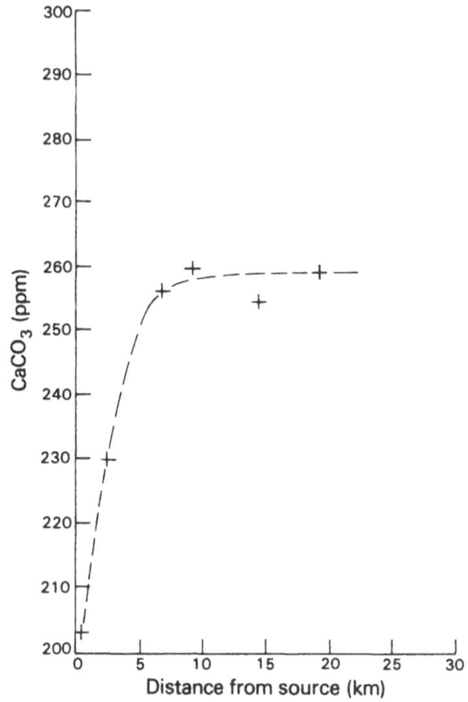

Figure 5 Graph showing relationship between calcium carbonate ($CaCO_3$) content of stream water and distance from stream source.

off to an equilibrium state at 260 ppm when saturation is reached. The graph shows this change and the establishment of equilibrium at about 10 km from the stream source.

Secondly, clear breaks in the slopes of the trend line indicate when thresholds occur in the relationship between two variables. A project measuring the relationship between the channel forms of a stream and its discharge at various points along its course found two separate relationships, which corresponded to a subtle change in the geology of the drainage basin. Unless the data had been graphed, such a threshold would not have been noticed and the statistical correlation coefficient of the variables would have been lower (Fig. 6).

The use of graphs as an aid to identifying anomalies has been discussed in Chapter 2. One normally looks for the points on a graph to be close to the regression line when testing hypotheses; but, as suggested earlier, those points that deviate far from the line are also of interest because they stimulate the development of new hypotheses or the refinement of the original one. It may be the deviants that one is looking for, however, and this third use of graphs may be illustrated by a project concerned with evidence of recent climatic change. As a

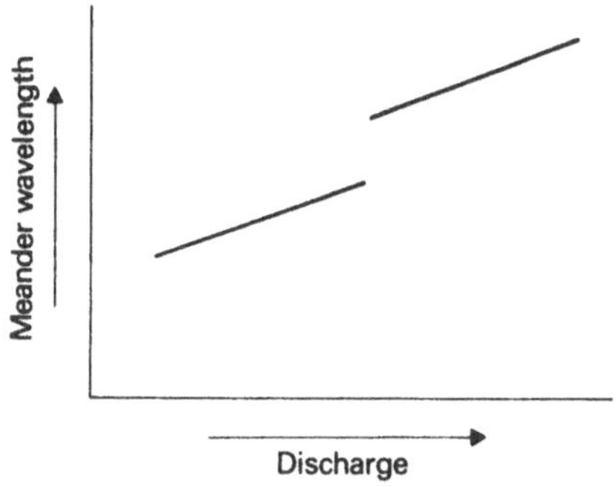

Figure 6 Graph showing a threshold in relationship between two variables.

result of winter storm damage, many beech trees were felled at various local sites during 1978. This gave an opportunity for measuring the width of the tree rings of the cut trunks in these different places. Tree ring widths reflect growth conditions. Sometimes wide rings are specific to an individual tree, as when rapid growth results from the felling of its competing neighbours. Narrow rings, which result from disease or branch lopping, may also be confined to one tree. But when all the trees have wide or narrow rings in particular years, the inference will be that general growth conditions have been unusually good or bad. The interpretation of tree ring widths is complicated by the fact that the rings tend to be narrower as the tree reaches maturity and they may be distorted if the tree is growing asymmetrically. In order to identify years with moist summers and long growing seasons, therefore, it is necessary to distinguish between those wide tree rings that are specific to one tree and those that are general, and to eliminate the variations in width that result from the normal growth cycle of a tree. Graphs enable this to be done and the exceptional years to be identified (Fig. 7). In this particular project the tree ring evidence was used to verify other types of evidence of past climates.

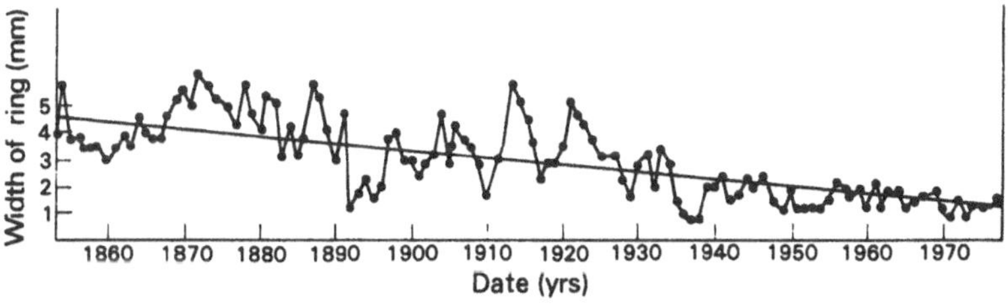

Figure 7 Graph showing variations in annual tree ring widths. This graph may be compared with the graphs for other trees to identify years when growth has been abnormally good or poor.

Fourth, graphs may be used for predictive purposes. When a trend line has been plotted, readings for Y may be interpolated from readings of X, and the prediction may be checked against reality, so again testing the relationship between the two variables.

For graphs that show a linear relationship, it is useful to fit the regression line. This may be done by eye, or accurately with the formula:

$$(y-\bar{y}) = r\frac{\sigma y}{\sigma x}(x-\bar{x})$$

where r is the product moment correlation coefficient (see p. 35), σy is the standard deviation of the y scores, σx is the standard deviation of the x scores, $\bar{y}$ is the mean of the y scores and $\bar{x}$ the mean of the x scores. Two separate y values can be calculated by substituting any two x values in the formula, the intercepts are plotted, and the straight line passing through these two points is the regression line. Where the readings on the X axis are at regular intervals, as with a time scale, a simple formula may be used. Statistics textbooks explain the latter procedure. The accurate fitting of a regression line is time consuming and although it gives polish to a study, for most projects where a correlation is clear the line drawn by eye will be adequate. The interpretation of the graph is much more important than the refinements of its construction.

One further comment may be made about the construction of graphs. Sometimes the readings on one or on both axes will show such a wide range that there may be a need to compress the scale. For example, when plotting the number of different urban services against the populations of urban centres, the population scale may need to be compressed in order to include the conurbations. Here a log scale may be used. It is possible to purchase semi-log and log graph paper for such purposes, but remember that with a semi-log or log-normal graph, a straight line trend may be, in fact, showing an exponential growth curve.

Statistical correlation

Having produced at least a rough graph of the relationship between the two variables for one or more of the above reasons, a statistical correlation technique may now be adopted to describe precisely the closeness of the relationship. The simplest method involves the use of the *Spearman rank* correlation formula:

$$r_s = 1 - \frac{6\Sigma d^2}{n(n^2-1)}$$

This is a non-parametric test, which means that there are no special requirements about the data sample used other than the obvious one that both sets of paired data can be ranked. Study 2 uses this technique, and the calculation and its layout are explained there. The project was concerned with testing the hypothesis that the orientation of Cotswold valleys corresponds to the orientation of joints in the limestone rock, and the technique involved the ranking of the lengths of valley and of the numbers of joints in each 10° sector.

Another simple correlation technique which is occasionally used in projects is the *point-biserial.* This applies when one set of data is arranged on an interval scale and the other is divided into two groups. A project investigating the spatial distribution of prices for fruit and vegetables in shops in Gloucester, used the technique in the following way. It was hypothesised that produce would increase in price with distance from the city centre where higher turnover and strong competition would reduce prices. This was tested with the Spearman rank correlation formula and a coefficient of +0·25 was calculated. Then the relationship between the rateable value of the shop and the prices of its goods was measured by the same method and a coefficient of +0·27 was found. Neither measurement was statistically significant. However, in the course of investigation it was noticed that some shops obtained their produce from wholesalers and others direct from growers, and the question was raised as to whether this had any bearing on prices. The point-biserial formula:

$$r = \frac{\bar{x}p - \bar{x}q}{\sigma t}\sqrt{pq}$$

enables this idea to be tested.

The prices of the sixteen shops were listed and price indices were calculated from a range of goods of comparable quality. From this list the standard deviation σt was obtained:

$$\sigma t = \sqrt{\frac{\Sigma\,(x - \bar{x})^2}{n}}$$

where $x - \bar{x}$ is the difference between the mean shop price $\bar{x}$ and the price of goods in a particular shop, and n is the number of shops. Then, the mean price for shops with wholesaler supplies was calculated $\bar{x}p$, as was the mean price for shops with direct grower supplies $\bar{x}q$. Finally, with p representing the proportion of shops with wholesaler supplies, and q the proportion with market supplies, the coefficient was obtained.

The calculation was set out as follows:

Shop number	*Price index*	$x-\bar{x}$	$(x-\bar{x})^2$	*Price index (wholesaler shops)*	*Price index (growers shops)*
		(98·0 − 118·0)			
1	98·0	−20·0	400		90·0
2	102·5	−15·5	240		102·5
3	103·0	−15·0	225		103·0
4	111·0	− 7·0	49		111·0
5	113·0	− 5·0	25		113·0
6	116·0	− 2·0	4	116·0	
7	116·5	− 1·5	2	116·5	
8	119·0	+ 1·0	1		119·0
9	120·0	+ 2·0	4	120·0	
10	121·0	+ 3·0	9	121·0	
11	121·5	+ 3·5	12	121·5	
12	121·5	+ 3·5	12	121·5	
13	123·0	+ 5·0	25		123·0
14	123·5	+ 5·5	30	123·5	
15	125·0	+ 7·0	49	125·0	
16	156·0	+38·0	1444	156·0	
Σx	1890·5	$\Sigma(x-\bar{x})^2$	2531	Σxp 1121·0	Σxq 769·5

$$\bar{x}=\frac{1890{\cdot}5}{16}=118{\cdot}0 \qquad \bar{x}p=\frac{1121}{9}=125{\cdot}0 \qquad \bar{x}q\ \frac{769{\cdot}5}{7}=110{\cdot}0$$

$$\sigma t=\sqrt{\frac{\Sigma(x-\bar{x})^2}{n}}=\sqrt{\frac{2531{\cdot}0}{16{\cdot}0}}=12{\cdot}6$$

Point-biserial coefficient:

$$r=\frac{\bar{x}p-\bar{x}q}{\sigma t}\sqrt{(pq)}=\frac{125{\cdot}0-110{\cdot}0}{12{\cdot}6}\sqrt{\left(\frac{9}{16}\times\frac{7}{16}\right)}\quad=+0{\cdot}59$$

Checking in significance tables the critical value of r(with $n=16$) is 0·425 at the 0·05 level, and 0·60 at the 0·01 level. The coefficient of 0·59 is, therefore, significant at the 0·05 level but not at 0·01. It was concluded that the source of supply of produce was a significant factor in determining vegetable prices, and the project then moved on to consider whether there were spatial factors influencing the type of supplier shops used.

The most powerful correlation technique is the *product moment.* This requires normally distributed data and involves a lengthier calculation. However, where numbers are manageable and where the requirements are met, this technique should always be used. It may be used, for example, when measuring the relationships between the various components of a river basin. Here such characteristics as the mean valley slope angles, the width and depth of valley incision, the width and depth of the river channel, the grade of bedload, and the bankfull discharge may be correlated one with the other. The product moment correlation formula is:

$$r=\frac{\frac{1}{n}\Sigma(a-\bar{a})\,(b-\bar{b})}{\sigma a\ \sigma b}$$

and the calculation is set out as follows where the relationship between width and depth of valley incision is being measured for a small drainage basin in the Forest of Dean.

Width of valley incision (a)	$(a-\bar{a})$	$(a-\bar{a})^2$	*Depth of incision* (b)	$(b-\bar{b})$	$(b-\bar{b})^2$	$(a-\bar{a})(b-\bar{b})$
240·0	+161·2	25985·4	25	+14·5	210·2	+2337·4
27·5	− 51·3	2631·7	8	− 2·5	6·2	+ 128·2
50·0	− 28·8	829·4	7	− 3·5	12·2	+ 100·8
80·0	+ 1·2	1·4	15	+ 4·5	20·2	+ 5·4
38·0	− 40·8	1664·6	5	− 5·5	30·2	+ 224·4
52·0	− 26·8	718·2	7	− 3·5	12·2	+ 93·8
59·0	− 19·8	392·0	8	− 2·5	6·2	+ 49·5
21·0	− 57·8	3340·8	5	− 5·5	30·2	+ 317·9
29·0	− 49·8	2480·0	6	− 4·5	20·2	+ 224·1
140·0	+ 61·2	3745·4	11	+ 0·5	0·2	+ 30·6
52·0	− 26·8	718·2	11	+ 0·5	0·2	− 13·4
7·0	− 71·8	5155·2	3	− 7·5	56·2	+ 538·5
66·0	− 12·8	163·8	7	− 3·5	12·2	+ 44·8
110·0	+ 31·2	973·4	15	+ 4·5	20·2	+ 140·4
210·0	+131·2	17213·4	25	+14·5	210·2	+1902·4
Σa 1181·5	$\Sigma(a-\bar{a})^2$	66013·0	Σb 158	$\Sigma(b-\bar{b})^2$	647·0	$\Sigma(a-\bar{a})(b-\bar{b})$ 6124·8

$$\bar{a}=\frac{1181{\cdot}5}{15}=78{\cdot}8 \quad \sigma a=\sqrt{\frac{66013}{15}}=66{\cdot}3 \quad \bar{b}=\frac{158}{15}10{\cdot}5 \quad \sigma b=\sqrt{\frac{647{\cdot}0}{15}}=6{\cdot}6$$

Substituting in the formula:

$$r = \frac{\frac{1}{n}\Sigma(a-\bar{a})(b-\bar{b})}{\sigma a \cdot \sigma b} = \frac{\frac{1}{15}\times 6124{\cdot}8}{66{\cdot}3\times 6{\cdot}6} = 0{\cdot}9$$

This is significant at the 0·01 level and indicates a very close relationship. When each variable has been correlated with every other one, the results may be set out as a correlation coefficient matrix. Using a *t* test or correlation tables, the significant relationships may be identified. Incidentally, this procedure is the basis for delimiting morphological systems.

All correlation tests should be followed by a test for significance, which shows the likelihood of a randomly selected pair of variables giving the same result. It is customary to reject the hypothetical relationship if the correlation coefficient is below that of the critical 0·05 level. (0·05 level means that there is a 5% chance that randomly selected pairs of variables would give as close a correlation as the one obtained.) Statistics texts will explain the methods of applying significance tests. It is also conventional to adopt the negative approach to statistical hypothesis testing by aiming to reject the null hypothesis. The null hypothesis states that there is no significant statistical relationship between the two sets of variables.

It must be stressed that correlations do not in themselves show cause and effect relationships. A correlation may result from both variables having a third variable as their cause, or there may be a long chain of relationships linking the two, or one may be a superordinate of the other, or the relationship may be by chance – although in this latter case an enlarged sample would alter the relationship. Further investigation is required before the relationship can be described as causal. Correlation merely shows covariance, i.e. the two phenomena vary together.

Chi square test

The *chi square test* χ^2, which has been described in Chapter 2, is also frequently used in hypothesis testing where one wishes to demonstrate that a particular distribution pattern varies with that of another factor, such as the numbers of fruit farms with slope aspect. Again, a null hypothesis is stated; that the number of fruit farms is not related to aspect of land. Then the expected number and the observed number for each aspect sector are substituted in the formula:

$$\chi^2 = \sum \frac{(O-E)^2}{E}$$

A chi square value greater than that shown in the chi square tables for the particular number of degrees of freedom and the confidence level selected would enable the null hypothesis to be rejected and the inference to be drawn that fruit farm distributions and aspect are related.

A form of the chi square test can also be used to measure the degree of association between two phenomena, and this is particularly useful in biogeographical projects where the association between two species is to be measured. Randomly located quadrats are set out along an environmental gradient, such as from the interior of a wood to the centre of a clearing, and the species of plant occurring in each quadrat are listed. Then, using a 2×2 contingency table, it is possible to calculate the expected frequency of the two species occurring together in the quadrats. The quadrats may be divided into four groups according to the species found in them as the table shows:

		Species *A* +	–	
Species B	+	both species (a)	only *B* (b)	($a+b$)
	–	only *A* (c)	neither (d)	($c+d$)
		($a+c$)	($b+d$)	n

The probability of quadrats with both species occurring in them (E), is given by the equation:

$$E = (a+b)(a+c)/n$$

This is the expected reading in the chi square formula and the number of quadrats in which both species are found is the observed reading. $(O-E)^2/E$ gives the chi square value. A species association matrix may be constructed showing the chi square values for the various associations of each species, and in this, the higher numbers show which species have the closest distribution ranges, either because of some relationship between the species, or because their tolerance limits coincide. The method is shown in Figure 8.

Table showing presence of plant species in 24 quadrats in Forest of Dean

Plant species	*Quadrats* (√ indicates species present)																						
	1	2	3	4	5	6	7	8	9	10	11	12	13	14	15	16	17	18	19	20	21	22	24
Pteridium aquilinum		√	√				√				√	√			√	√		√	√	√	√	√	√
Rubus sp.	√	√	√	√	√	√	√	√	√			√	√	√	√		√		√	√	√	√	√
Lonicera periclymenum	√	√		√				√															
Epilobium angustifolium			√				√		√	√	√										√	√	√
Hedera helix	√			√									√	√					√				
Quercus robur	√			√		√		√					√	√			√		√		√		
Hypericum pulchrum							√			√					√								
Endymion non scriptus	√					√											√						
Dryopteris filix-mas	√			√										√									
Gramineae sp.					√		√	√	√						√				√				

The association between, for example, *Pteridium aquilinum* (bracken) and *Quercus robur* (oak), is measured as follows:

Observed frequency, i.e. number of quadrats in which both species occur, is 2.
Expected frequency $(a+b)(a+c)/n$ is 5·25 This is obtained from the matrix.

Species *A* *Pteridium*

		+	−	
Species *B* *Quercus*	+	*a* 2	*b* 7	*a*+*b* 9
	−	*c* 12	*d* 3	*c*+*d* 15
		a+*c* 14	*b*+*d* 10	*n* 24

$$\frac{9 \times 14}{24} = 5{\cdot}25$$

Chi square value $\frac{(O-E)^2}{E}$ is $\frac{(2-5{\cdot}25)^2}{5{\cdot}25} = 2{\cdot}12$

The association matrix would then be constructed as below:

	Pteridium	*Rubus*	*Lonicera*	*Epilobium*	*Hedera*	*Quercus*	→
Pteridium	x					2·12	
Rubus		x					
Lonicera			x				
Epilobium				x			
Hedera					x	5·18	
Quercus	2·12				5·18	x	
↓							

Figure 8 The production of a species association matrix.

Nearest neighbour analysis

Spatial patterns may be described by the technique of nearest neighbour analysis. The method has been widely criticised in recent years and some of its snags are revealed in Study 4, where the calculations do not confirm the expected and visual pattern of rural settlements at different dates. It remains, however, the best available technique for project work where measurements of the spacing of geographical phenomena are required. The method of calculating the nearest neighbour index from the formula:

$$R_n = \bar{D}\text{ obs}/\bar{D}\text{ ran}$$

is explained in Study 4. Here an adaptation of the formula for linear patterns is considered.

A project studying the distribution pattern of shops in a city street may require an objective measurement of the clustering or dispersal of shops selling particular types of goods. An examination of the shop patterns of constrasting towns may also require a similar measurement. Here the modified nearest neighbour analysis formula is used:

$$R_n = \frac{\bar{D}\text{ obs}}{0{\cdot}5\left(\dfrac{L}{n-1}\right)}$$

where $\bar{D}$ obs is the mean number of premises between nearest neighbours of shops of a particular type, L is the total number of premises on the side of the street being investigated, and n is the number of shops of the type being studied. Consider, for example, the shops in the Promenade, Cheltenham, where in 1978, the list included three jewellers, ten ladies clothes shops and eight shoe shops. Nearest neighbour indices were for jewellers:

$$\frac{19}{0{\cdot}5\left(\dfrac{45}{3-1}\right)} = 1{\cdot}69$$

for ladies clothes:

$$\frac{2{\cdot}2}{0{\cdot}5\sqrt{\dfrac{45}{10-1}}} = 0.88$$

and for shoe shops:

$$\frac{3{\cdot}75}{0{\cdot}5\sqrt{\dfrac{45}{8-1}}} = 1{\cdot}17$$

and show the strong 'comparison' factor for ladies clothes. The pattern has not always been the same and, working through Kelly's directories and other records of earlier years, it can be shown how the shop pattern of the street has changed – though always with an emphasis on the fashionable and the luxurious.

Hypothesised changes in street shops patterns through changes in fashion, the invention of new products, the nationalisation of enterprises, or increased competition for prime sites, may all be tested by this linear modification of nearest neighbour analysis.

Mann-Whitney U test

In the course of an investigation it may be necessary to determine whether or not two samples of data belong to the same or to different sets of data, or to see if there is a significant difference between two sets. The most powerful method of analysis for this problem is the one-way analysis of variance, but a very simple way of testing is to use the Mann–Whitney U test. This is illustrated in the following project on scenery evaluation.

Several techniques of measuring the scenic quality of landscapes have been devised, and one of the earliest is that of Linton who classified the quality of Scottish scenery primarily on the basis of vigour of relief and variety of landuse. If the technique is as objective as is claimed, it should be possible for different observers to assign identical scores to each unit of landscape area. Fieldwork in the Nantlle Valley in North Wales gave an opportunity for such an experiment to be carried out with sixteen observers independently applying Linton's scheme of scoring scenic quality on a kilometre grid basis. When the scores for each square had been collected their standard deviation was calculated in order to show which types of scenery gave the widest variation in scoring. From discussion about the difficulties of scoring, it was hypothesised that observers moving down the valley into areas of less impressive scenery would score less generously than observers working up the valley into the more spectacular landscapes. To test this the Mann-Whitney test was used.

The total scores given by each observer to the whole valley were arranged in rank order as follows, with A referring to up valley scores, and B to down valley scores. 140(A), 141(A), 142(A), 147(A), 152(A), 155(B), 157(B), 159(A), 165(B), 167(B), 167(A), 174(B), 185(B), 185(B), 191(B). The number in the smaller group is referred to as n_1 and in the larger n_2, but in this case with equal numbers of investigators in A and B groups $n_1 = n_2 = 8$. The U statistic is obtained by summing the number of observers in group A that precede each B, or the number of B's that precede each A – whichever is the lowest total. In this example, it is the number of B's preceding A's, and is $0+0+0+0+0+0+2+3=5$. Referring to Mann-Whitney test tables with $n_2 = 8$, $n_1 = 8$ and $U = 5$, the probability is read off at 0.001. The null hypothesis that there is no significant difference between the two groups of scores is rejected at both 0·05 and 0·01 levels because the probability reading is below both, and the conclusion is that there is a statistically significant difference between the scores up and down the valley. But this is not in the direction hypothesised because up valley scores are significantly lower than down valley. The explanation is probably that the exact location where observations were made in each square differed with the two approach routes. This remains to be tested in the field.

Variance

Many tests mentioned earlier involve the measurement of the standard deviation of scores. The square of the standard deviation or *variance* $\Sigma(x-\bar{x})^2/n$, may also be used to summarise data in an objective way. Several of our projects have examined the textbook description of regions, and earlier reference has been made to examples which tackle such questions as, 'Can one recognise the traditional *pays* of the Paris Basin from land use patterns?' or, 'Does the land use of Belgium correspond to the textbook's regional divisions?' In these studies the textbook description has been compared with data collected from land use transects recorded from a minibus en route for a more distant field course. A wide variety of crops may be recorded, some occupying extensive areas, others just a single plot, and most somewhere in between. To make a textbook generalisation about the land use of an area will inevitably mean that some of these crops have to be ignored – but which? The group of crops giving the lowest variance is the most accurate summary of land use and it is calculated as follows. A ten kilometre transect in Beauce in 1975 gave readings of 31·2% barley, 24·1% maize, 23·7% wheat, 11·3% haricots, 3·2% sugar beet, 1·1% oats, 0·8% peas, and smaller stretches of several other crops. For a two-crop combination, the accurate percentage would be 50% barley and 50% maize. So the variance for a two-crop combination is:

$$\frac{\Sigma(x-\bar{x})^2}{n} = \frac{(31{\cdot}2-50)^2+(24{\cdot}1-50)^2}{2} = 512{\cdot}1$$

with 33·3% for each of the three leading crops the variance for a three crop combination is:

$$\frac{(31{\cdot}2-33{\cdot}3)^2+(24{\cdot}1-33{\cdot}3)^2+(23{\cdot}7-33{\cdot}3)^2}{3} = 60{\cdot}4$$

and for four crops it is:

$$\frac{(31{\cdot}2-25)^2+(24{\cdot}1-25)^2+(23{\cdot}7-25)^2+(11{\cdot}3-25)^2}{4} = 57{\cdot}2$$

The variance for five crops is 102·8, and for six 128·0. The lowest variance is for four crops and so the most accurate summary of landuse in this ten kilometre stretch is barely, maize, wheat and haricots. Notice that data on labour requirements, or crop values, or tonnage of yield would place the emphasis on a very different combination of crops, but the same technique could be used to summarise the data.

Other techniques

Some projects use techniques that are specifically related to a particular type of study, such as using Reilly's Law in the calculation of break points between settlements to delimit service areas in order to compare these with field work data; or the calculation of a central location for scattered sites using weighted co-ordinates in order to find the optimum site, for example, of a fire station based on one year's calls; or the measurement of the connectivity between places by network analysis to test the effect of a new road as in Study 7. Many of these techniques are explained in the human geography texts such as Bradford and Kent (1977), Tidswell (1976), or Everson and Fitzgerald (1969).

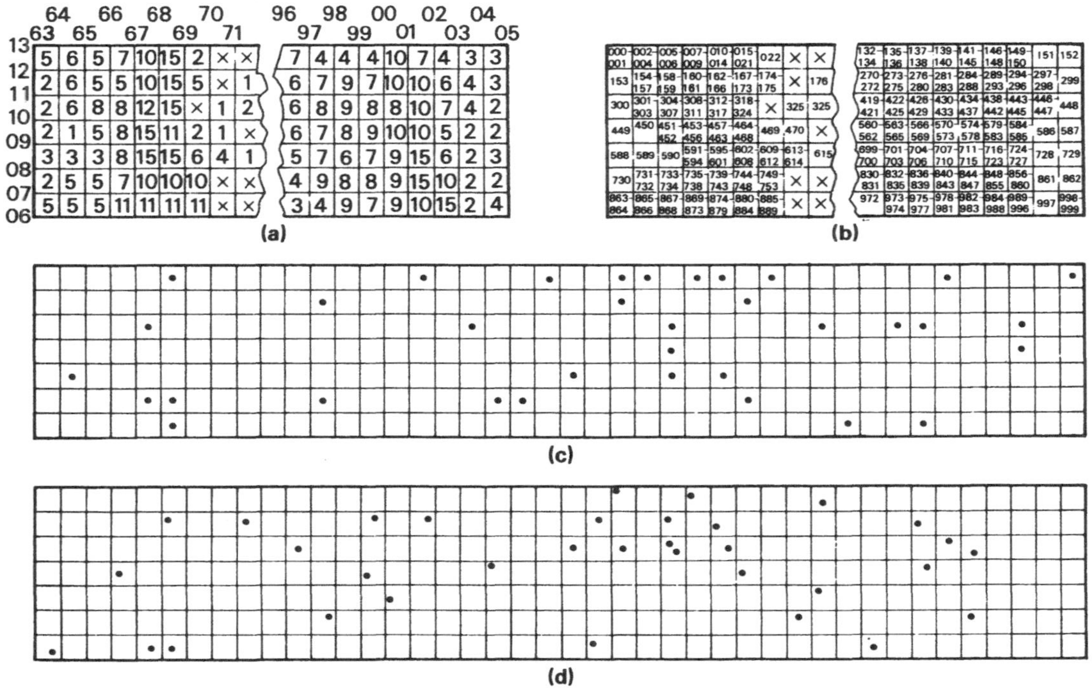

Figure 9 Simulation of a distribution pattern of country houses. (a) Each grid square is given a score which is the sum of the values it is judged to possess on the following scales: views 0–10, nearness to water 0–10, nearness to London road 0–5, land between 30 m and 150 m 0–3; (b) scores are adjusted to range from 000 in top left square to 999 in bottom right square; (c) random number tables are used to give readings of 35 locations; (d) actual country house locations for comparison.

It must not be assumed that hypotheses may only be tested statistically. Sometimes, a simulated distribution pattern based on hypotheses may be produced to compare with the real one. The study of country house distributions mentioned in Chapter 3 was tested by ranking the location factors which had been identified, and after giving them values based on their likely importance, an adjacent area was surveyed for possible country house sites. The predicted sites were then checked with the actual country house locations, so testing the original hypotheses (Fig. 9).

The visual comparison of distributions through the use of map overlays may also prove effective as a method of testing where the 'cause' and 'effect' are seen to coincide. This is demonstrated in Study 3 where sound waves are shown to travel with relief features in some weather conditions. Similarly, the finding of anticipated evidence to verify a hypothesis as in Study 1 may prove sufficient.

Statistical testing of hypotheses is not absolutely essential and the tests are only to be regarded as tools. They are, however, extremely valuable tools because they give a measure of the quality of the hypotheses.

As the scope of geography projects increases and as new types of problems arise so new methods of analysis and of testing will have to be devised. The techniques briefly reviewed in this chapter concern problems that have been treated in a static sense, and as the analysis of trends and of changing patterns figure increasingly in geographical investigation, so a new range of methods will be necessary. Some schools have access to computer terminals and this too will increase the sophistication of statistical analysis, but high quality work can be performed without these expensive aids, and most valuable results come from experimenting with different methods of treating the data with the simple aids of a pencil and a scrap of paper.

6 Conclusion and presentation

When each hypothesis has been satisfactorily tested, the conclusion may be written. Frequently, the results of these tests form the main items of the conclusion, but there is more to the conclusion than a mere summary of the results and a drawing together of the various strands of inquiry.

In the conclusion one attempts to answer the question set in the early stages of the project. Some answers will be firm and precise, as when correlations have been shown to be statistically significant or when predictions have been proved accurate. Others will be tentative, as when a sample may be too small to be unquestionably representative (as in Study 3) or when relationships are not clear. But whichever is the case, a simple concise statement in answer to the question should be incorporated in the conclusion. Just as the type of question considered in projects varies, so too will the kind of conclusion, but the two will be related, and even at this late stage it may be necessary to make slight adjustments to the wording of the question so that it fits exactly the terms of the conclusion. This is not a desirable practice but it may avoid a misfit conclusion.

When the question asks, 'What factors determine a particular distribution pattern, or what are the reasons for the development of a particular landscape feature?', the conclusion may go beyond the basic list of factors or reasons which have been disclosed in the study. An attempt may be made to rank them in order of importance, so indicating a deeper understanding of the problem. An example of this is given on page 44.

The conclusion should also relate the project's findings to a broader scene than the local one if this is possible. Comments may be made on the implications of the results for other areas, or other times, and in this way the position of the investigation within the whole field of geography may be made clear. Study 1 illustrates this point.

Many projects fall short of the initial expectations. Perhaps a technique of analysis chosen in good faith has failed to yield the anticipated results, or the exact data required for processing were unobtainable and cruder information had to be substituted, or there was insufficient time to develop the argument as far as one would have liked. Where these weaknesses occur, the results should be treated with appropriate reservation. With the sense of relief and satisfaction that comes on completing the work, there is always a danger of over-reacting. But be confident to draw attention to the strengths of the study and do not undersell the work by being hypercritical. The balanced, logical and coherent argument that has characterised the project so far should be continued in the conclusion.

As the project has developed, new ideas will have arisen. Sometimes they are of such importance that the whole project could have been centred on them, but they came too late. These may be mentioned in the conclusion. Research is continuous and although the limits may have been reached through shortage of time or resources, new vistas of inquiry may have been opened during the course of study, and in the conclusion brief attention may be given to them.

The studies that follow demonstrate the variety of conclusions that are possible. In some the hypothesis testing has pointed clearly to a solution to the problem, in others every hypothesis has been rejected and progress has only been made through a process of elimination, and in others there is a mere glimmer of light but deficiency of data or of technique has prevented much illumination. Rather than dwelling longer on the general points of the conclusion, reference may now be made to these examples where the more specific points are illustrated and discussed.

Finally a few practical hints are given on the writing and presentation of the project.

The examination boards publish regulations about the submission of projects. There will be a word limit and candidates may be penalised for exceeding the limit. This limit is fixed in order to help the candidate, and is judged to be appropriate to the length of time available for the project and to the level of study. Weaker candidates tend to exceed the limit through failure to reject the irrelevant and through a lack of clear focus in the study.

Most boards also advise that work should be neatly bound together and enclosed in simple folders which, apart from being cheap, have the added merit of a lighter weight for postage than do heavy ring files.

The effort involved in the project warrants careful presentation and a neat, legible, handwritten or typed and corrected script is necessary.

Sometimes, an abstract of two or three hundred words which summarises the whole project is recommended in addition to the normal introduction and conclusion. Subdivision of the work helps in the actual writing process, which is best tackled bit by bit. Each separate section can be written and then left for a while before checking to see if it is simple and clear.

When an interview is to be given, reference to the report is made easier if each page is numbered.

Important sources of facts and ideas should be given at the end of the study. All work relies to some extent on the work of others, and such help should be acknowledged. The quality of your own work is enhanced by this because it shows that time and effort have been given to considering the background to the specific problem treated in the project. The same applies to the sources of information for maps and diagrams.

Maps, diagrams, photographs and tables of figures should be included in the final report only if they are relevant. They should be integrated with the text, and reference should be made to them. Just as one takes care with the choice of statistical tests so one should take care with the choice of ways of presenting data in graphical form. Generally, maps are preferable because they indicate that some considerations of spatial patterns have been made. Much time can be wasted in copying ornate maps. Slavish reproduction does nothing to improve the study particularly when such essentials as scale and orientation are often overlooked.

It may be possible to look at the work of candidates in previous years and from other schools, and so to note the qualities to emulate or avoid. This is probably the simplest way of appreciating the main points of presentation.

EXAMPLE PROJECTS

Study 1 The origins of the minor landforms of a scarp slope

The project's background: field work in the Wye Valley near Monmouth

A very important part of the A-level course in geomorphology is the study of slopes, within which the measurement of slope angles and the properties of the materials of which they are composed are usually studied. A useful preliminary field work exercise is to measure the maximum slope angles on different rock exposures and to correlate these measurements with such geological variables as rock hardness or degree of permeability. When there is a common erosional agent and the degree of exposure to this agent is more or less uniform, as in the incised valley meanders of the Wye, these correlations can be instructive. Such field work leads to a more perceptive observation of slopes and to an easier recognition of such significant features as breaks in profile. The smooth profiles of some textbook diagrams are soon regarded as generalisations, which may be far from reality.

Forming the question

The scarp slope of the Cotswolds has a very uneven surface, which consists of numerous low mounds or hummocks of various shapes and sizes – rather like coastal sand dunes in form, or the earthworks of deserted medieval villages (Fig. 10). Such a surface came as a surprise to one accustomed to the textbook diagrams, and the conflict between the expected form and the observed form led to the question tackled in this study: *How may these scarp slope hummocks have been formed?* Other interesting geographical questions might have been asked concerning, for example, their distribution, their form, or the processes operating on them today, but the chosen problem involves these others to some extent and leads to a type of investigation which differs from the approaches that the other questions would have required.

Hypothesis generation from reading

The review of the available literature on the subject was not very helpful. The standard geomorphology texts gave some ideas but

Figure 10 Low hummocks on the Cotswold scarp slope.

nothing specifically on these landforms. A chapter in W. Dreghorn's *Geology explained in the Severn Vale and Cotswolds* (1967, Newton Abbot: David & Charles), contained a reference to glacial periods when, 'the subsoil remained frozen to great depths and the frost fractured rocks slid down hill when the upper layers thawed out', but little more. A paper by R.J. Chandler (1971) in *Slopes: form and process* (IBG Special Publication) referred to the knolls and hummocks on the scarp face in Northamptonshire as outliers or the products of rotational slumping, but their dimensions seemed rather larger than the Cotswold features considered here. Solifluction, rotational slumping and outliers then were possible explanations to be considered.

Hypothesis generation from observation

Human activity could not be ruled out as a potential cause because, although the hummocks are found above the limits of ridge and furrow, the area has a long history of earthworks. At many sites along the escarpment are promontory Iron Age forts and adjacent to the area of study (GR 929164) are the remains of the Crickley Hill fort with its ramparts. Quarrying for freestone has also taken place along the top of the escarpment. Recent quarry waste is easily identified, but the use of local stone extends back over many centuries and waste dating from the earlier periods may not be so clearly recognisable.

So it was apparent that several explanations could be considered as hypotheses for the formation of the hummocks and that others might arise as investigation proceeded. Different causes might be expected to produce slightly different forms and structures, and therefore prior to excavation the expected forms resulting from each hypothetical process were delineated as clearly as possible.

Hypothesis elaboration

Ramparts of local Iron Age camps have been excavated and diagrams of sections cut through them have been published. Those at Crickley Hill show piles of stones covered by a thin layer of soil and have some indications of regularity in construction. They may also show the reddish marks of burning, and their siting is normally near a hill top.

Since quarrying was for freestone, *quarry waste* would probably have been of this material and would consist of stones of unrequired sizes or possessing some property which made them difficult to cut. The accumulation of waste would be disorderly and would be located below and close to a quarry face. Quarrying marks might be preserved on the stones.

Landforms resulting from *rotational slumping* would consist of disturbed material or of blocks sloping back towards the scarp. A slump scar would have existed on the uphill side and might be traced by excavation if it had been in-filled by soil creep. One might also expect to find evidence of contemporary slumping associated with periods of heavy rain, or of snow melt, and with the issuing of spring water.

An *outlier* would consist of more resistant material *in situ*, and since the location of the hummocks is below the outcrop of the inferior oolite, their structure would be of lias clay or marlstone, although some weathered oolite might have accumulated on the uphill side.

If the cause of the hummocks was *solifluction* in periglacial conditions, they would probably contain frost shattered blocks of oolite which had slid downhill on a slippery surface and evidence of that surface would exist below the blocks. Subsequent soil forming processes and soil creep would have smoothed the outline of the blocks.

Deductive reasoning led to the suggestion of these distinguishing marks for each hypothesis, and hypothesis testing consisted of the careful search for these items of evidence.

Hypothesis testing

There is a large number of hummocks, and the two chosen for

excavation were selected partly because of their accessibility, and partly for their manageability with pick and shovel and for their contrasted forms. A more ambitious excavation programme would have required a sampling procedure to ensure that a range of sizes, forms and locations was considered. Permission to dig was sought from the landowner.

The most economical dig was a trench wide enough to work in, and about a metre deep, cut at right angles to the long axis of the hummock, beginning at a point about a metre away from the edge of the hummock on the uphill side and continuing to the same distance below the lower edge. In both hummocks a large block of oolite was found. The dimensions of these blocks were determined by digging and by probing with a crowbar, and found to be 11·5m × 14·6m × 3m in the first hummock, and 13·1m × 4·6m × 3·1m in the second. Their shapes varied and so did their surfaces, some surfaces had a clean edge but most were severely weathered. Both blocks were easily related to the outcrops above but some of the weathered rock fragments in the hummocks were less easily identified and were probably not related to the blocks. The overall shapes of the hummocks that were surveyed (Fig. 11), seemed to be determined primarily by the shapes of the blocks and their angles of rest. Other interesting discoveries from the dug sections concerned the material surrounding the blocks. This was clearly stratified with two well defined soil horizons covering the blocks and several others which had accumulated on the uphill sides (Fig. 12). Samples from these horizons were analysed for their matrix size, organic content and $CaCO_3$ content in order to identify them.

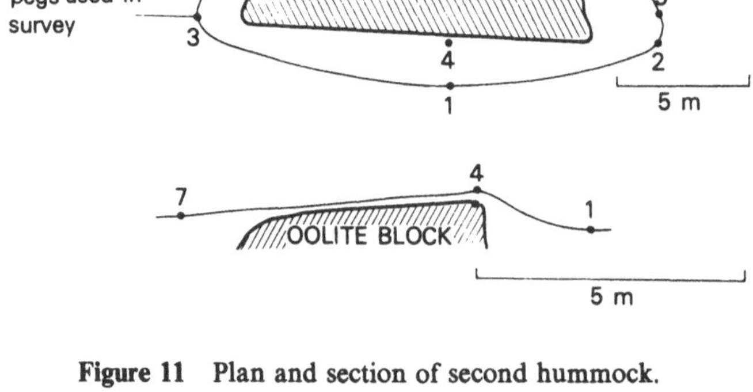

Figure 11 Plan and section of second hummock.

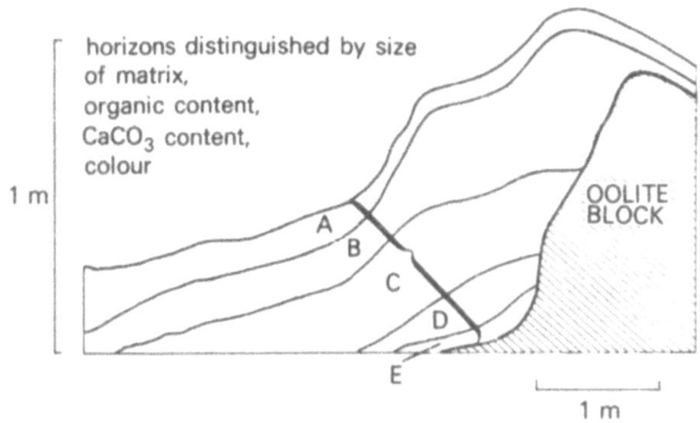

Figure 12 Section of upper part of first hummock.

With reference to the hypotheses: the sections were very different from those through Iron Age ramparts; such large blocks of freestone were unlikely to be rejected as quarry waste and no pick markings were found; the clearly defined horizons suggested that slope processes had been so slow that the layers were not disturbed or that major movements occurred long ago and subsequent conditions have been stable. Beneath the blocks was a layer of yellowish clay so the hummocks could not be regarded as outliers of oolite. No evidence of slump scars was found here.

A further control section was then dug on a smooth part of the scarp face close to the excavated hummocks. This showed the same two horizons that capped the oolite blocks and below these was a clay horizon similar to that on which the blocks rested. The analysis of these horizons gave results similar to those mentioned above.

Conclusion and comment

The conclusion reached, therefore, was that the hummocks were composed of blocks of oolite, which had slid over this clay surface from the rock outcrops above, and on stabilisation the blocks had allowed weathered material to accumulate behind them. Normal soil-forming processes had given the smooth outlines to the present day hummocks. It was proposed that solifluction in periglacial conditions was the most likely explanation.

The fact that several hypotheses were advanced as explanations of the hummocks, and that one was shown to have more support than the others, does not mean that all other hummocks have this solifluction origin. Further investigation of different-sized hummocks and in different positions along the scarp face would be necessary before such a conclusion could be reached. They may be examples of equifinality, i.e. similar landforms that have developed in different ways from different origins.

It was suggested that future study might also consider the blocks of quartz conglomerate that litter the slopes bordering the Forest of Dean as phenomena of similar origin, and that mapping the distribution of these Cotswold hummocks with particular reference to slope angles and altitude would shed more light on their formation.

Study 1 illustrates a simple, clearly defined research exercise, involving the formulation of several hypotheses, careful thought about their likely outcome, and a programme of methodical testing. Its conclusions are suitably reserved. Many similar studies could be envisaged dealing with small-scale land forms like these, which are manageable for one person.

Study 2 Structural influences on valley orientation

Drainage patterns have always attracted a good deal of attention from geographers. Some patterns are associated with one particular type of landform and others are related to different landforms. The Cotswold dip slope might be expected to show a simple dendritic pattern. But a brief scanning of OS sheet 163, 1:50 000 is sufficient to identify a much more angular design than this. The courses of the rivers Churn, Coln, and Windrush (tributaries of the upper Thames), have north–south sections alternating with north west–south east sections and many of their feeder valleys follow the same orientations. How can one explain this unexpected pattern?

The study outlined below was undertaken after an Easter field course in South Devon which included work on the origin of Dartmoor tors. The elegant paper by D.L. Linton (1955) The problem of tors, *Geog. J.* had been read and had been compared with the findings of J. Palmer and R.A. Neilson. Points of similarity and of conflict between the two theories of tor formation were noted and the field work exercises included a search for supporting evidence. In both accounts the jointing of granite is given prominence, and the frequency of jointing is thought to be a dominant factor in the breakdown of granite whether by physical or chemical processes or by both. The shapes of tors are also joint controlled and in his study of Dartmoor landforms, R.S. Waters has widened the scope of joint control to include the succession of basin and gorge, tor and col – basins corresponding to intersecting planes of closely spaced joints and tors corresponding to widely spaced joints in these planes.

With some background knowledge, then, of the role of jointing in landform development, of the conspicuous jointing of the Cotswold oolitic limestone, and of the angular drainage pattern of Cotswold streams, the precise question to be answered was quickly identified: *Can jointing explain the orientation of Cotswold valleys?*

Some ideas were gleaned from the standard geomorphology texts but the most helpful reference was the paper by E.H. Brown (1969) Jointing, Aspect and Orientation of Scarp Face Dry Valleys, near Ivinghoe, Buckinghamshire, *Trans. IBG.* This paper provided information on the methods of investigation and also encouragement as to the feasibility of the study.

Data collection

To assess the relationship between jointing and valley orientation, measurements of the two variables were required. A sample area within which these data were to be collected could have been chosen at random, but for ease of access and for abundance of information an area was selected which included large scarp quarries and also much of the Churn Valley – bounded by grid lines 92, 02, 10 and 20.

Valley orientations were plotted from an OS 2½in. sheet, which with 25ft (7·5m) contour intervals shows more detailed relief information than the other available sheets, and these were checked in the field. Straight lines were drawn along each linear section of the valley. Dry valleys were included and no distinction was made between the different valley sizes contained within the sample area. Figure 13 illustrates the type of map produced.

It was now possible to measure the length of valleys lying along each compass bearing, and by summing them to identify the most important valley orientations. The semicircular frequency diagram Figure 14 shows the results. The length along each compass bearing corresponds to the total of valley lengths in that direction. Plotting was in 10°

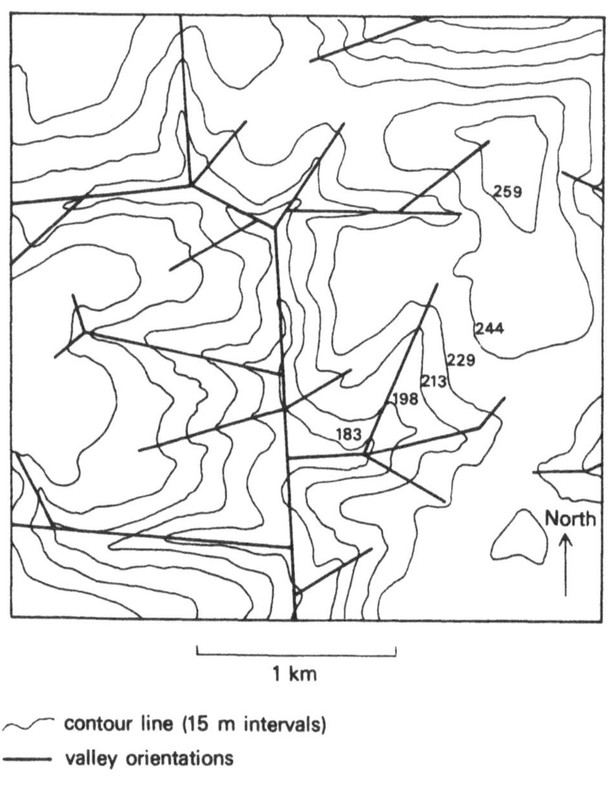

Figure 13 Map showing valley orientations of a sample Cotswold area.

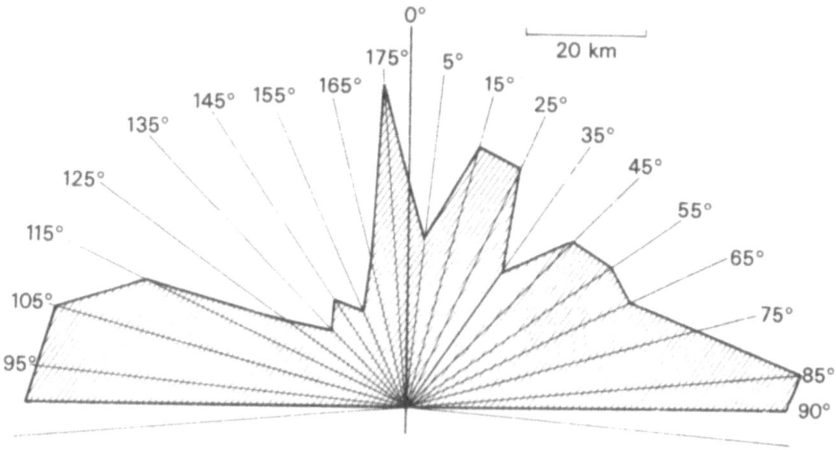

Figure 14 Valley orientation frequency diagram.

sectors. Peaks occur at 15°, 85° and 175°, the east–west orientation being particularly pronounced. The question of whether or not these peaks are statistically significant will be dealt with later.

Measurement of *joint orientations* necessitated visits to exposed rock faces. In the Cotswolds, where quarrying has been a traditional rural activity for centuries, exposures of the oolitic limestone are common, although many quarries are now overgrown. Some quarries are at sites where local disturbances of structure, through cambering and slipping, have occurred, and these too were avoided. Out of a possible thirty-one recorded quarries in the area, the following five were found suitable for examination and each had at least fifty clearly visible joint planes: Leckhampton (GR 949185 and GR 947184), Crickley (GR 929162), Climperwell (GR 929114) and Highwayman (GR 962115).

A joint was arbitrarily defined as a clearly visible fissure crossing three or more bedding planes, and its orientation was measured by taking the compass bearing of a brass rod driven into the joint plane. Because of the deep embayment of the quarries it was possible to obtain joint orientation readings over almost 360° at these sites and, in all, about five hundred readings were taken. The number of joints in each 10° sector was also plotted on semicircular frequency diagrams (orientations having been adjusted from magnetic to grid north to allow comparison with the valley orientation diagram).

As work progressed, it became apparent that the well marked peaks and troughs of joint frequency on these diagrams were being more or less repeated from one quarry to another; some quarries had three peaks, others four, but the orientations were similar. This suggested that the joint orientations recorded at the quarry sites could be applied to the whole area of study. Also, that as both inferior and great oolite were exposed in the quarry sample, the same joint patterns were applicable to both strata. Thus a river flowing over inferior oolite in its upper stretches and great oolite in its lower would probably cross rocks with similar joint orientations.

Figure 15 shows the joint orientation frequency for the combined quarry readings. With peaks at 55°, 85°, 115° and especially 165° it does not correspond well with the valley orientation diagram. The visual comparison was not thought to be sufficient, however, and a statistical correlation technique was used to describe the relationship between joints and valleys.

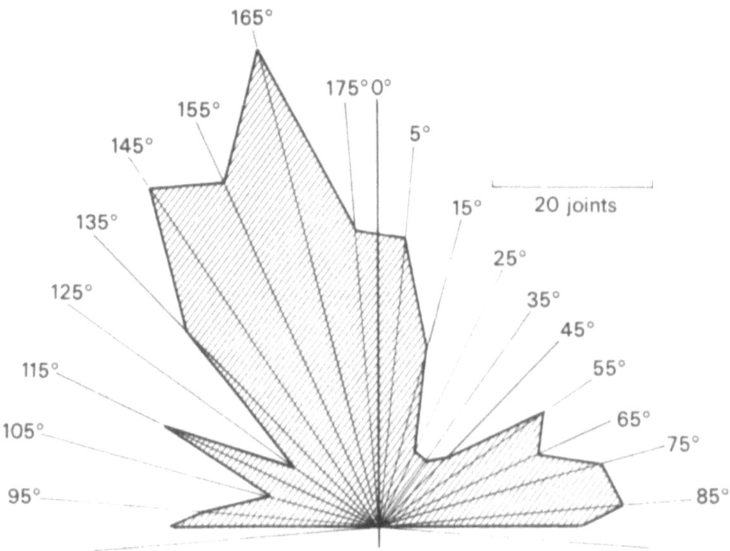

Figure 15 Joint orientation frequency diagram.

Correlation of valley and joint orientations

An inspection of the distribution of joint frequency readings and of valley length totals showed that neither could be regarded as normally distributed and so a parametric method of correlation, i.e. the product moment correlation, could not be applied, although the data were recorded on an interval scale. It was therefore more appropriate to use the Spearman rank correlation method. The data and calculations are set out in the table below.

Spearman rank correlation of length of valley and number of joints per 10° sector.

Sector (°)	*No. of joints*	*Rank*	*Length of valley (km)*	*Rank*	*Difference in rank (d)*	d^2
1– 10	36	5·0	28·5	12·0	−7·0	49·00
11– 20	23	11·0	44·5	7·0	4·0	16·00
21– 30	10	17·5	43·5	8·0	9·5	90·25
31– 40	10	17·5	27·5	13·0	4·5	20·25
41– 50	12	16·0	39·0	11·0	5.0	25·00
51– 60	25	10·0	41·0	9·5	0.5	0·25
61– 70	22	12·5	41·0	9·5	3·0	9·00
71– 80	29	8·0	50·0	5·0	3·0	9·00
81– 90	31	7·0	67·0	1·0	6·0	36·00
91–100	22	12·5	63·0	2·0	10·5	110·25
101–110	14	14·0	61·5	3·0	11·0	121·00
111–120	28	9·0	48·5	6·0	3·0	9·00
121–130	13	15·0	24·0	15·0	0·0	0·00
131–140	34	6·0	17·5	17·0	−11·0	121·00
141–150	51	2·0	21·5	16·0	−14·0	196·00
151–160	47	3·0	17·0	18·0	−15·0	225·00
161–170	61	1·0	24·5	14·0	−13·0	169·00
171–180	37	4·0	53·5	4·0	0·0	0·00
						Σd^2 1200·00

Substituting in the correlation statistic:

$$r_s = 1 - \frac{6\Sigma d^2}{n(n^2-1)} \qquad r_s = 1 - \frac{6 \times 1206}{18(324-1)}$$

$$= 1 - 1{\cdot}245$$
$$= -0{\cdot}245$$

This value of r_s is not significant at the 0·05 level and it was concluded that there was no significant statistical correlation between joint and valley orientations. The inference drawn was that the orientations of these Cotswold valleys could not be attributed to the influence of joint patterns on erosion. It should be noted, however, that even if a significant correlation had been shown to exist between joint orientations and valley orientations, one could not have concluded on this evidence alone that jointing caused the valley directions. Correlation does not prove a cause and effect relationship.

Further inquiry

In this case the absence of a significant correlation raised two new questions: why does such a well developed joint system have no apparent effect on valley orientation and can one formulate other hypotheses as possible explanations for the unusual valley pattern?

Three suggestions were made in answer to the first question. Either the valleys may have been cut in circumstances when jointing did not exert an influence on erosion, e.g. in permafrost conditions, or the joints are so close to one another that differential erosion has not occurred, or other influences may have outweighed those of joints.

The study area is crossed by a series of *fault lines* running roughly east–west. Many of these fault lines correspond with valleys, and in fact a total length of 141 km of valley overlies fault lines. Assuming that the geological map is accurate, this means that 19·7% of all valleys coincide with faults, and faulting must, therefore, be considered as a major factor in influencing valley orientation.

It is conceivable that faulting may have masked the effects of joints. To check this, all valleys which may be attributed to faults were excluded from the data and the rank correlation exercise was repeated. Again, no significant correlation was found.

Another structural influence on drainage direction is the *dip* of the rocks. The regional dip here is to the south east but local dip varies considerably. In the north east of the area the geological map shows the direction of dip varying between 190° and 205°, and in the south west it is between 130° and 165°. Valley orientations in these areas (excluding the possible fault line valleys) show peaks at 195° and 205° in the north east and at 135° in the south west – again suggesting that dip should be considered as a factor in valley orientation.

The study was concluded at this stage with the comment that the control of valley orientation is complex and that faulting and dip appeared to be more significant than jointing, though neither could be shown to account for more than a small proportion of valley directions.

Comment

It will have been noticed that the study only considered the linear dimensions of valleys – other dimensions such as width and depth could have been included. Valleys of all widths and depths of incision were given equal weighting and the study could have looked at the relationships between joints and valleys of streams of the different stream orders. This point is important because a difference has already been disclosed between the major valleys with their north–south, north west–south east sections and the most significant valley orientations which were east–west.

Logically, a χ^2 test should have been given to both valley and joint orientation distributions before the investigation of relationships had been developed, because it is necessary to show that the variations of frequency of both joints and valleys between the different compass sectors are statistically significant and not likely to be the result of chance.

Maps of the area's geology showing rock type, faulting, dip and quarry sites, valley orientations, and drainage pattern, were included in the study.

Apart from illustrating a particular type of geomorphological study and the use of a rank correlation technique, this study also shows that a project which fails to demonstrate an anticipated result can still have equal merit to one in which the expected results are confirmed.

Study 3 An analysis of sound patterns

In this example a leisure interest in bell ringing gave the impetus for a novel local study.

Optimum locations for such varied phenomena as industrial establishments, rural settlements, farm buildings, and post boxes are frequently calculated in local studies and Study 4 considers some of the problems encountered with this type of project. Normally the oldest and traditionally most important building of a rural settlement is the parish church, and its precise location within the parish is also worth investigation. Historically, the call to worship or the summoning of parishioners for other purposes was by church bells, and the question posed in this study was: *Are churches sited where their bells can be heard throughout their parish?*

Sound is a form of wave motion and, although the sixth-former working on this problem was not a physicist, an elementary understanding of the principles of sound was clearly necessary before useful hypotheses could be formulated and tested. Reflection and refraction occur with sound as with all wave forms and in the latter case air temperature differences may be the cause. Sound waves travel more rapidly in warm air than in cold, so the distance sound travels near the surface will be influenced by the temperature lapse rate of the air. As the diagram Figure 16 shows, a temperature inversion will cause sounds to carry long distances as the sound waves are refracted outwards, while a steep lapse rate and the consequent upward refraction of sound waves causes reduced audibility ranges at surface levels.

Similarly, the refraction of sound waves can be caused by the vertical wind gradient. As wind speeds are checked by friction near the earth's surface, surface winds have a lower velocity than those at higher levels. This leads to the bending outwards of sound waves in a leeward direction and so to greater audibility, and upwards on the windward side giving reduced audibility, again shown in the diagram.

Some suggestions have been made that humidity variations may contribute to variations in audibility, and in particular that sound travels further in foggy weather, but as humidity varies with temperature and atmospheric turbulence, it is difficult to relate audibility to any one atmospheric variable.

Hypotheses

Following these preliminary comments on the movement of sound waves a set of hypotheses was advanced.

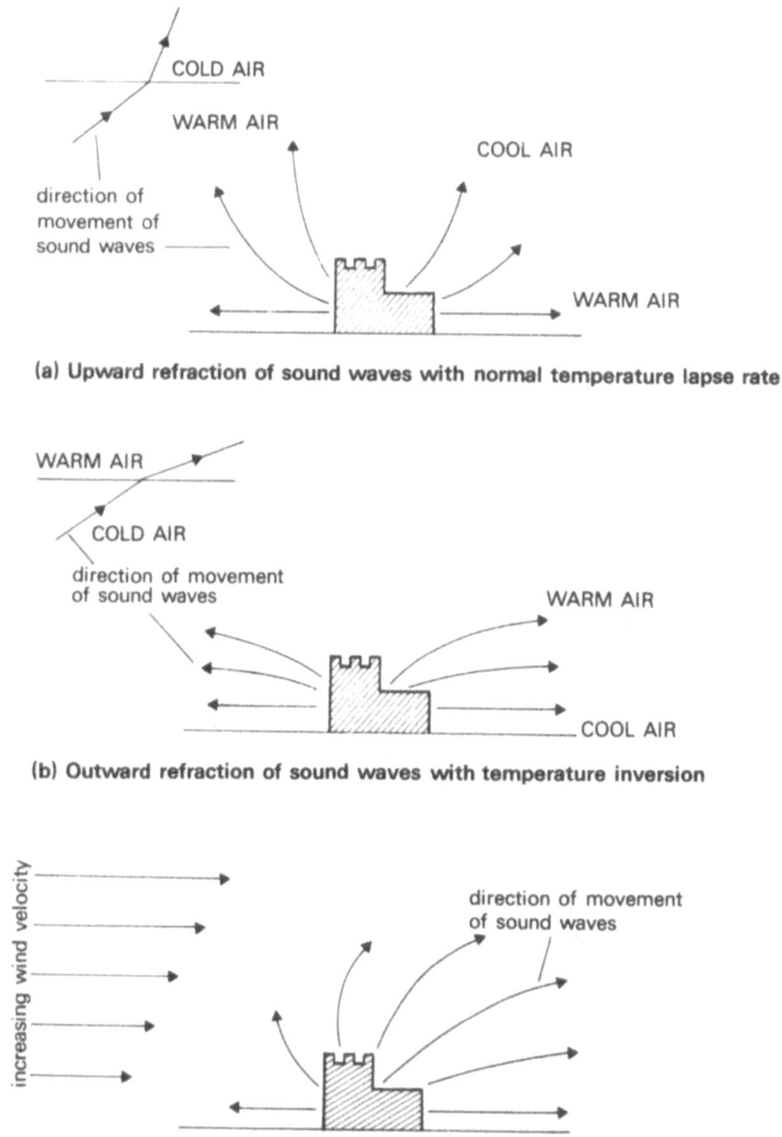

(a) Upward refraction of sound waves with normal temperature lapse rate

(b) Outward refraction of sound waves with temperature inversion

Figure 16 Diagrams showing possible causes of refraction and distortion of sound waves.

(a) To enable bells to be heard throughout the parish a central location was selected for the church.
(b) The sound emitted by bells in a tall tower travels further since friction with the surface is reduced and so large parishes have tall belfries.
(c) The higher the site of the church relative to the surrounding relief the further the sound travels.
(d) Relief distorts lines of equal audibility (the term **isophone** was coined).
(e) Cyclonic conditions muffle sound and so reduce audibility.
(f) Anticyclonic conditions, and particularly when temperature inversion occurs, are associated with long distance sound ranges.

Data collection

The collection of data to test these hypotheses required careful planning. First, only churches with regular ringing practices could be used for most tests and this meant that sixty-six of the one hundred and seventy towers with bells in the Gloucester diocese were available for sampling. Secondly, for comparative studies, only rings (the collective term for all bells in a church) of equal weight could be used. Equal weights should give similar sounds. The weight is expressed in terms of the heaviest bell, the tenor, and these varied between 4·5cwt and 29·5cwt. The commonest weights were in the 10–12 cwt range and so the sample was now reduced to this group of twenty-one. Thirdly, a variety of sites was required and this constraint, together with that of accessibility, narrowed the sample to six. Six is too small a sample for statistical analysis.

Since three of the selected towers held bell ringing practices on the same evening of the week and it was normally impossible to take more than one set of readings in the hour and a half practice, the data collection was spread over several months. The requirement of specific weather conditions was an additional constraint on data collection.

The table below gives details of the sample towers, their sites and the weather and dates of readings.

In planning for the *measurement of audibility,* eight concentric circles at 1cm intervals (250m) were drawn around each church on OS 1:25 000 sheets and fifteen points were marked where roads crossed these circles – at least one point on each circle. During each set of readings every point was visited and the sound of bells was recorded on a portable tape recorder. An aural assessment of the volume of sound was also made at each point and this was particularly important towards the limits of the range of the bells where a trained ear is more responsive than a tape recorder. Later the tapes were played and the sound measured on a home made sound meter scaled 0–15.

Church	*Weight of Ring (cwt)*	*Situation*	*Rain*	*Sun*	*Calm*	*Wind*	*Good Visibility*	*Fog*	*Cloudless*	*8/8 Cloud*
Barnwood (GR 858177)	10·75	urban		Sep. 8	Nov. 17	Aug. 25	Sep. 22	Feb. 2	Oct. 6	Dec. 1
Maisemore (GR 813216)	11·00	rural	Sep. 2	Oct. 20	Sep. 9	Dec. 23	Oct. 28	Feb. 9	Sep. 16	Nov. 3
Churchdown (GR 882191)	11·5	hilltop	Sep. 1	Sep. 21	Sep. 18	Sep. 7	Sep. 29		Dec. 4	Jan. 8
Winchcombe (GR 022282)	12·00	valley			Oct. 16	Sep. 4	Aug. 28		Dec. 11	Oct. 23
Ashchurch (GR 927334)	12·00	flat			Dec. 14	Mar. 2	Nov. 9		Nov. 16	Jan. 26
Chedworth GR 052121)	10·00	undulating	Aug. 25		Oct. 17	Jan. 23	Oct. 24	Nov. 7		Jan. 30

The weather data including temperature; relative humidity; pressure; cloud cover, type and height; visibility; and wind direction and speed, were obtained from the local meteorological office and where possible were checked at the site.

Charts of the sound patterns recorded from each set of readings were produced and isophones were drawn in the same way as with any isopleth map. Figure 17a shows the chart for Chedworth on January 23.

Hypothesis testing

It was now possible to test the hypotheses.

(a) The central location within the parish was found by marking the centre of gravity of a cut out cardboard outline of the parish. The cardboard outline was suspended successively from two different points near its edge and on each occasion a plumb line was used to mark the vertical line which passes through the point of suspension and the centre of gravity. These two lines intersect at the central location. (It should be noted that ecclesiastical parishes, delimited on the 1840–5 tithe maps, were used in this study as they are closer to the original parish than the present day civil parishes.)

The distance between the calculated central location and the actual church location was then measured. A sample of six churches, chosen for other reasons, is too small for any significant conclusions but it was noted that this distance generally increased with increase in parish size, and that the chosen location for the church was likely to be at the most accessible site near the centre of the parish, or on the nearest prominent relief feature to this central point.

(b) The height of the bells above ground level was plotted against the size of the parish. The expectation was that the two variables would increase together but the graph (Fig. 18) shows, with one exception, an inverse relationship. Again, the sample is too small for statistical significance but the results do suggest that other factors determine belfry heights; perhaps the quality of building materials, the architectural style, the wealth of the church, or the firmness of the foundations outweigh audibility as a determinant of tower height.

(c) However, the effective height of the belfry is made up of the altitude of the church site in relation to its surroundings as well as the height of the tower. Thus a tall tower in a valley may have a much lower effective height than a short tower on a hill top. This may partly explain the surprising results in (b) for it was noticed that the height of the belfry varies inversely with the relative altitude of the church. So the combined effect of altitude of site and height of tower may

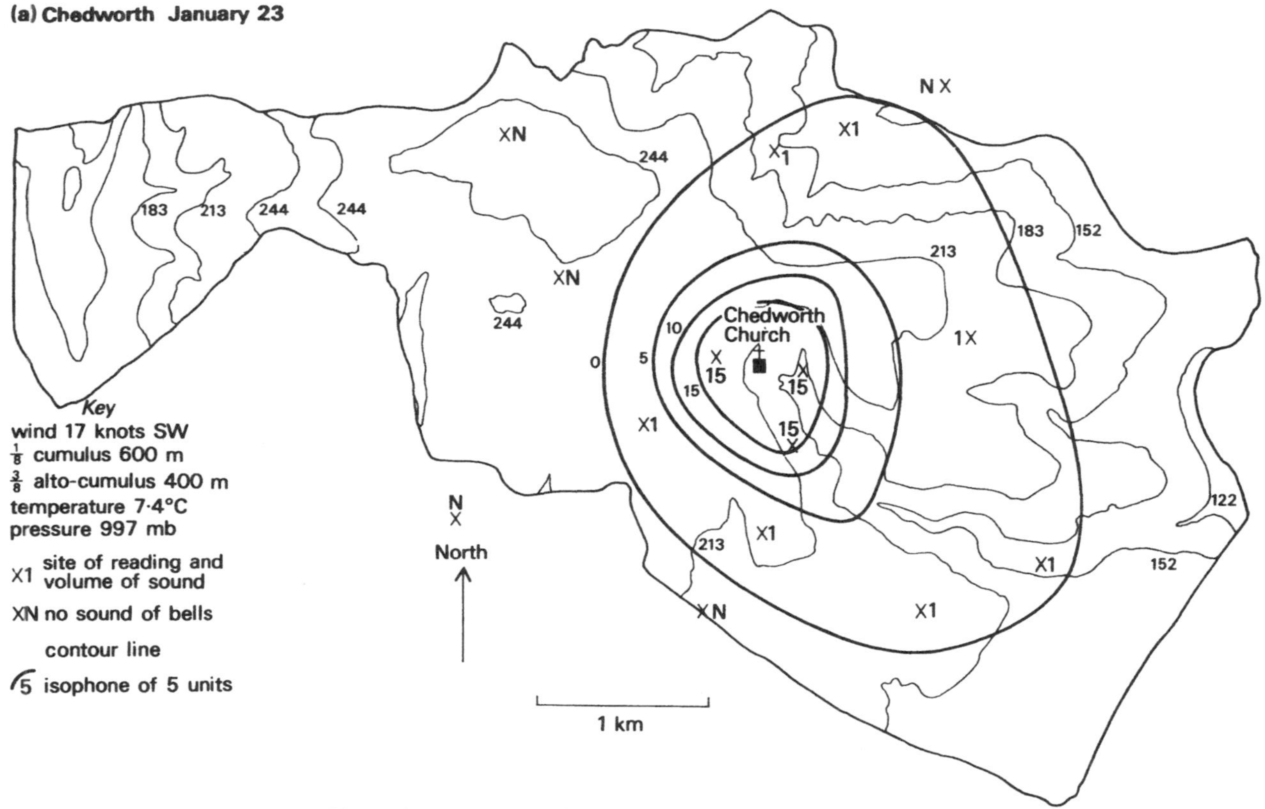

Figure 17 Effect of relief on audibility of church bells.

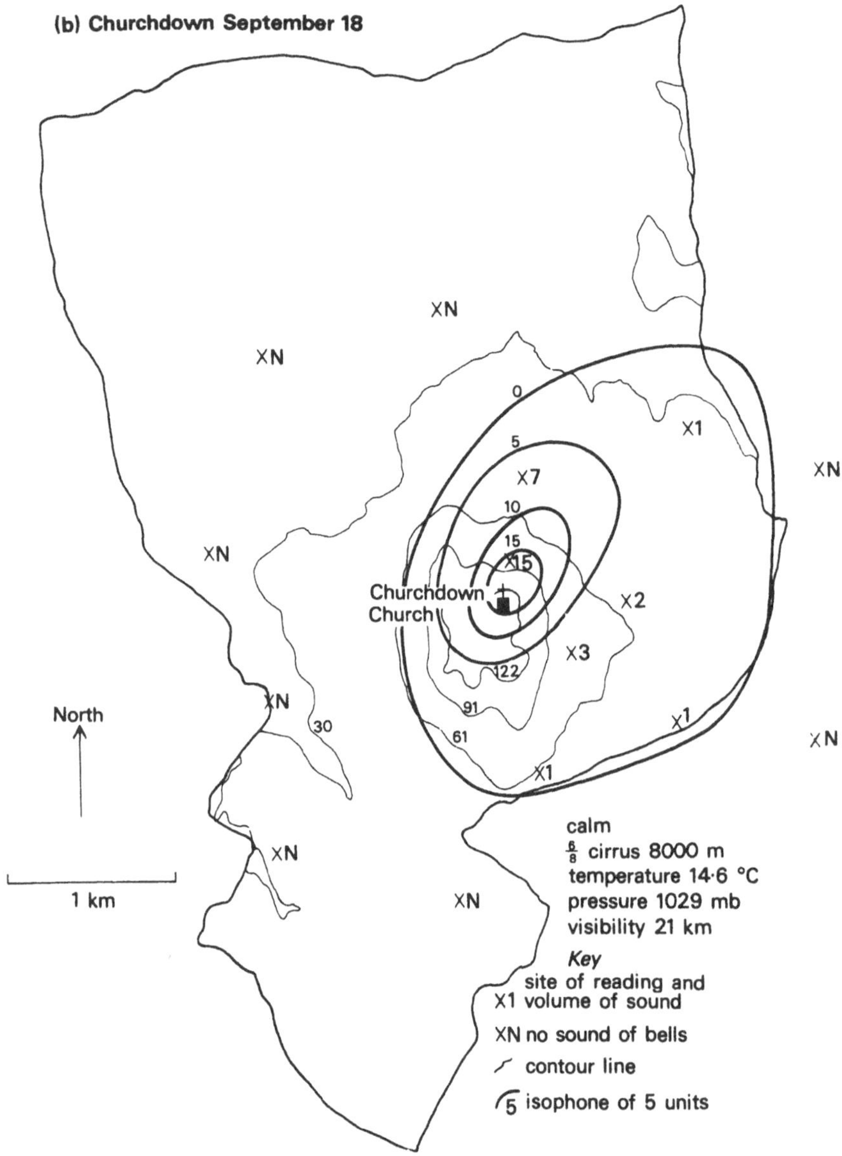

Figure 17 (continued)

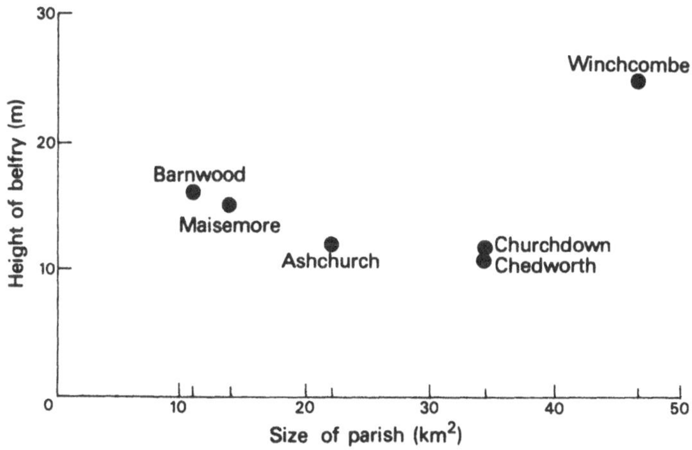

Figure 18 Graph showing relationship between size of parish and height of belfry.

show a positive correlation with parish size. A sample of 20–30 churches would be needed for this test.

(d) The analysis of several sound charts was needed to test whether or not relief distorts the movement of sound waves. For example, Figure 17a shows the isophone pattern at Chedworth with a moderate breeze blowing from the south west. Audibility is reduced, as might be expected, to the south west and the sound carries further to the north east, but the greatest distance the sound travels is to the south east and this corresponds to the line of the Chedworth Valley. Again, in calm conditions at Churchdown (Fig. 17b) the asymmetrical pattern may be explained by the fact that the church is located on the north eastern edge of a small flat-topped hill and hollows channel the sound down to the east and north. The charts show that sound does not travel so easily or so far uphill. It was concluded that relief features may distort sound patterns.

With regard to roughness of surface, the expectation was that a smooth surface with reduced friction would ease the passage of sound whereas a rough surface, for example, an urban landscape, would hinder it. Maisemore and Barnwood were selected to test this, and the clearest example occurred at Barnwood on a calm November day (Fig. 19). Here the sound of bells carried farthest to the south west, which corresponds to a wide open space. Elsewhere, buildings and trees surround the church.

(e) The effects of relief and surface roughness are not seen, however, in all charts, as weather conditions override those of topography. An unsolved problem arose here in that it was not possible to isolate the effects of a particular weather parameter such as relative humidity

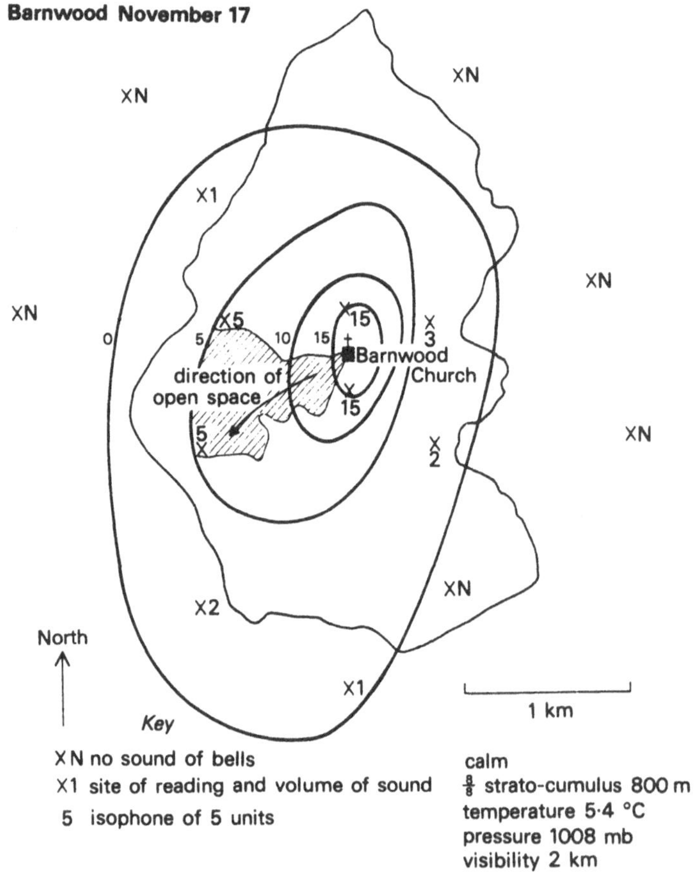

Figure 19 The effect of open space on the audibility of bells.

or wind speed or rainfall intensity. The hypothesis was therefore stated in terms of general cyclonic conditions.

Many charts showed the effect of wind by the reduction of audibility to windward. The comparison of the charts for one church with calm, light breeze and strong breeze conditions revealed that the greatest audibility range occurred with the light breeze – both calm and windy conditions muffle sound. Other charts led to a similar conclusion, and greatest audibility range was found with wind speeds between 1 and 10 knots. Part of the problem with strong winds is the increase in noise from twigs and leaves which competes with the sound of bells.

A muffling effect was also observed with continuous heavy rain. The volume of sound decreases rapidly with increased distance from the church (Fig. 20). No effect was noticed with variations in cloud cover. Cyclonic conditions, then, tend to reduce audibility, and strong winds and rain are the likely causes.

(f) With reference to anticyclonic conditions, a similar association of meteorological phenomena occurs. But on each occasion when a temperature inversion occurred near ground level the bells could be heard throughout the parish (Fig. 21). Sometimes fog was present on these dates, but not always; sometimes conditions were calm, but again not always. Temperature inversion seemed to be a strong factor contributing to the long distances sound travels in anticyclonic conditions.

In the course of examining the charts it was also noticed that the bells could rarely be heard beyond the traffic noise of motorways.

Conclusion

The response then to the question, 'Are churches sited where their bells can be heard throughout their parish?' was that it is possible to hear the bells throughout the parish but not in every kind of weather. With strong winds and rain the sound may not travel more than 750m in any direction and in these conditions no single site could be effective. The prevailing wind frequently limits audibility to the south west.

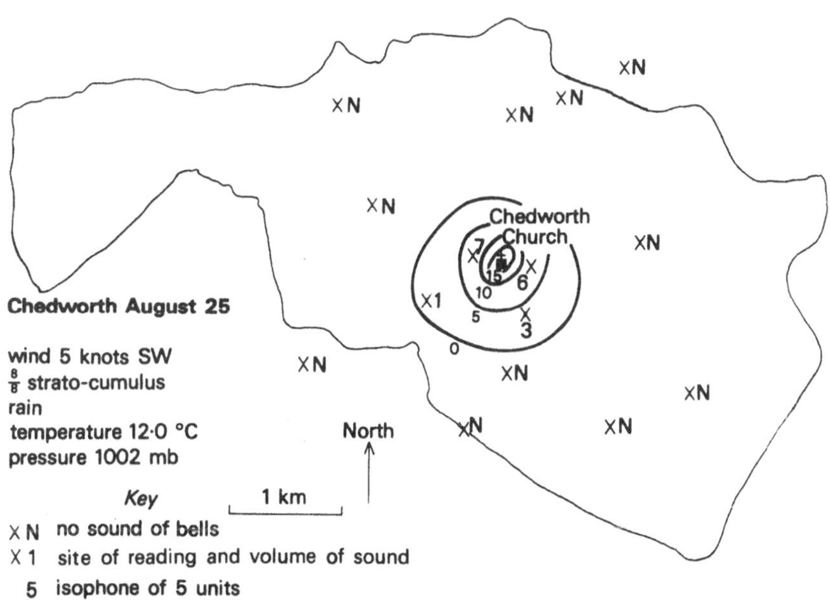

Figure 20 The effect of rain on the audibility of bells.

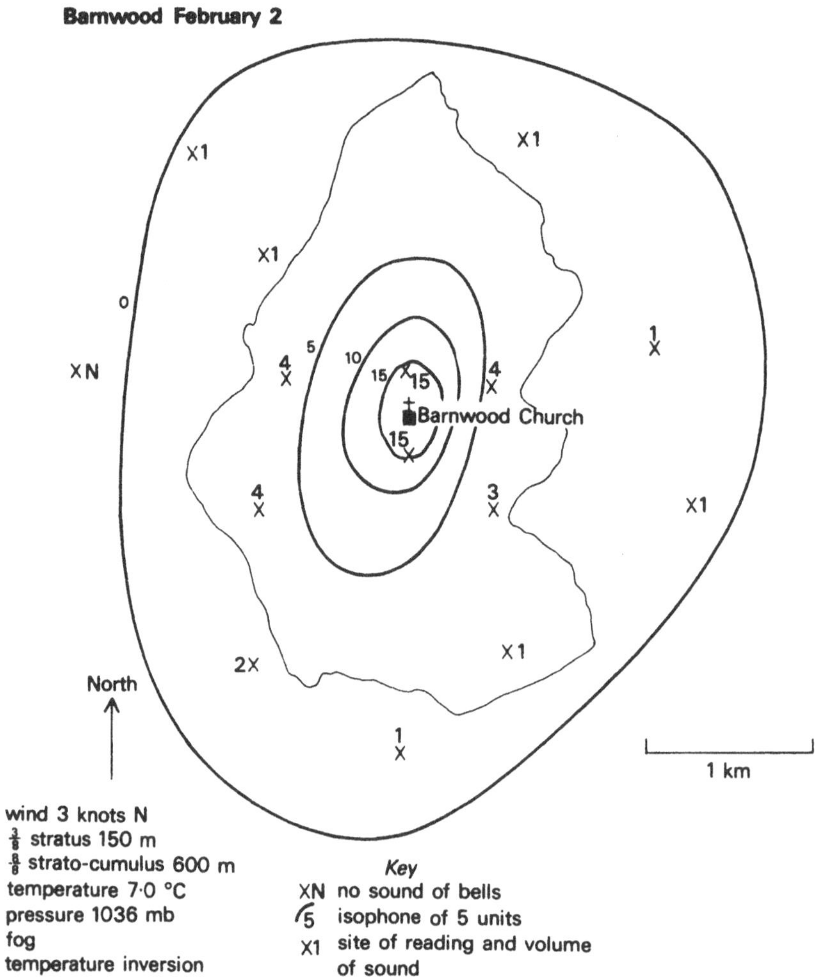

Figure 21 The effect of temperature inversion on the audibility of bells.

Comment

This study has been included among the examples for several reasons. First, it illustrates the geographical analysis of an environmental factor, sound, of which man is increasingly conscious. Aircraft noise, road traffic noise, industrial noise, crowd noise, all cause environmental problems to many people and some understanding of the way sound carries is necessary before any remedial action can be taken.

Secondly, the study shows how, with persistence, a person can conduct quite an impressive investigation almost single-handed. Many avenues for further study were disclosed but not followed. A larger sample would have been necessary for these, and to pursue any one of them would have resulted in curtailing the study elsewhere.

The inability to isolate specific meteorological phenomena for investigation was a problem for morale, but this is common to many similar studies involving co-variables.

Some may feel the study is tangential to geography, but it is encompassed by most definitions of the subject and the candidate was certainly not penalised by the examiners.

Again it exemplifies the thoughtful formulation of hypotheses and a carefully planned strategy to test them. One criticism may be made of the first hypothesis. This would only apply if bells were used at the time of the building of the church, which is unlikely.

Study 4 Local changes in rural settlement patterns

The study of settlement locations may be tackled in several ways and Studies 4 and 5 illustrate two of these approaches. In this case, a sample of settlements is examined in order to identify the settlers' site requirements that are inferred from the characteristics of the chosen locations, and in the second case a theoretical distribution pattern of settlements is developed from basic economic principles and this pattern is then checked against reality. The first illustrates an inductive approach, the second is deductive.

Every rural area in Britain has a history of settlement change, and the present day landscape may show the results of several superimposed settlement patterns, each pattern relating to the particular set of requirements of the time. Of course, some sites have advantages for settlement which have been recognised by successive colonists, and settlements at these locations have survived for long periods, but other sites have been used by one group of people and have then been abandoned when requirements have changed. To develop this theme of settlement evolution, a sample study area must have a number of dated settlements of different periods, and a terrain which varies in the opportunities it offers to rural life. The area selected here was circular, of 6 miles (9·6km) radius, round the Cotswold village of Hazleton (GR 0818). It was, therefore, of manageable size for field work and contained traces of Neolithic, Iron Age, Roman and Saxon settlement patterns (most of which are marked on OS sheet 163, 1:50 000). The scarp lands of England have many areas like this and similar work could also be attempted in upland Britain – though dealing with different groups of settlers. The title of the study: *How and why has the evaluation of the siting factors of rural settlement changed with time?* is along the lines of an A-level question and some of the ideas considered here come from normal A-level reading and in particular from M. Chisholm's *Rural settlement and land use* (1962).

The programme of investigation began with the mapping of the known settlement sites for each of the four periods. Although the OS sheet contained most of the information, it was necessary to check and supplement these details with records of more recent discovery contained in the archaeological journals of the county records office and especially in the *Transactions of the Bristol and Gloucestershire Archaeological Society*. After this study had been completed the Royal Commission on Historical Monuments published its report on *Iron*

Age and Romano-British Monuments in the Gloucestershire Cotswolds (1976) and some minor adjustment to the study has been necessary. Details of Saxon settlements were obtained from *The Place-Names of Gloucestershire.*

Neolithic settlement

The actual sites of Neolithic settlement are rarely known and none has been identified within the study area. It was assumed that the distribution pattern of long barrows, the principal monuments of the Neolithic people, would correspond to the distribution pattern of their settlements, so that the Neolithic settlement requirements could be inferred from the site factors of the barrows.

Before attempting to identify these site factors, however, it was important to show that there was some degree of spatial ordering of long barrows and that their distribution pattern was not random. An evenly spaced distribution would suggest, perhaps, equal size territorial units, and a clustered pattern might suggest a clustering of site advantages. Nearest neighbour analysis was chosen to describe the distribution pattern. This technique relates the mean distance between nearest neighbours, long barrows in this case, to the mean distance that would be expected if the pattern was a random one. An index value (R_n) of 1 indicates a random pattern (which does not necessarily mean that chance factors explain it), values approaching 0 show clustering, and a perfect regular distribution has an index value of 2·15. There are several snags with nearest neighbour analysis with respect to minimum sample size, to fitting boundaries to the area of analysis, and to the effect of paired distributions, but the technique was used here because the sample was thought to be large enough, the area was part of a larger region where similar distribution occurred, and most sites were discrete. The formula is:

$$R_n = \frac{\bar{D}\text{ obs}}{\bar{D}\text{ ran}} = \frac{\text{Measured mean distance between long barrows and their nearest neighbour}}{\text{Calculated mean distance if the distribution is random}}$$

$\bar{D}$ ran is given by the formula:

$$\bar{D}\text{ ran} = 0{\cdot}5\sqrt{\left(\frac{A}{N}\right)}$$

where A is the area of study and N is the number of long barrows. (Note formula for $\bar{D}$ ran, which is often misquoted.) The index value in this pattern is 1·48 (Fig. 22a).

Some indication of site requirements is given by the artefacts and bones found within the burial chambers of the long barrows, so the excavation reports were searched for this information. Bones of cattle, sheep and pigs, with fragments of grain storage jars, and an absence of hunting weapons and bones of wild animals indicate a pastoral economy with some cultivation, and thus a requirement for easily cleared, well drained land.

Each site was then visited, and the environmental factors that might have been advantageous to Neolithic people were recorded. All long barrows were on commanding sites, though not necessarily on the

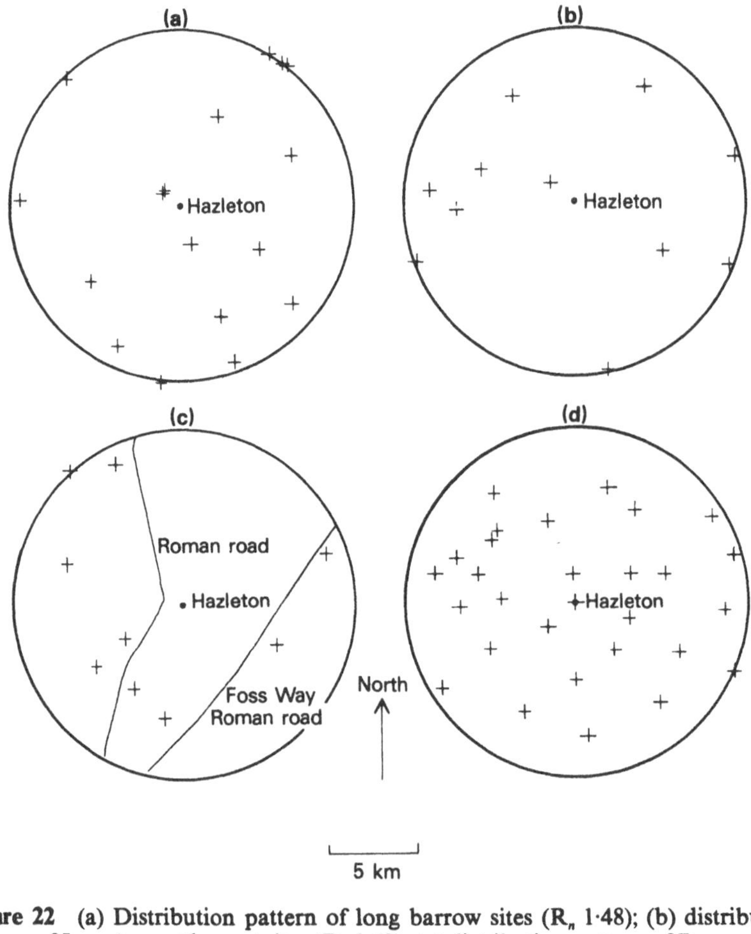

Figure 22 (a) Distribution pattern of long barrow sites (R_n 1·48); (b) distribution pattern of Iron Age settlement sites (R_n 1·62); (c) distribution pattern of Roman villa sites (R_n 1·21); (d) distribution pattern of Saxon settlement sites (R_n 1·37).

highest points. Some intervisibility of long barrows was noticed, and the possibility that the long barrow was sited so that it could be seen from all points of a Neolithic territorial unit was considered. The shapes of these hypothetical territories, however, make this unlikely. The distance to the nearest water supply was measured. Several long barrows were close to springs and the distance never exceeded a half mile 0·8km), but a half mile across difficult terrain may seem a long way to carry water and perhaps the settlement sites were nearer to the water sources. Spring sites may have been higher four thousand years ago. Soil is generally thin, stony and well drained, and the heavier clays of the valleys were avoided.

Although speculation about the significant east–west orientation of the long axis of long barrows is often made, the orientations in this sample were not regular, and a closer link appeared to be with the contour lines. Presumably construction of a long barrow with its long axis parallel to the contours is simpler. H.P.R. Finberg's interesting idea that the Neolithic settlement may be beneath the present day nearby village was also considered. With a ratio of fourteen long barrows to twenty-nine villages in the study area, it was noticed that in seven cases the line of the long axis neatly bisected the angle between the long barrow and the two nearest villages.

Iron Age settlement

A similar procedure of checking all the published archaeological records and then making observations in the field was followed for the settlement sites of the Iron Age and Roman periods. The line of promontory forts that follows the Cotswold edge and the important Iron Age settlement of Bagendon (GR 0106) lie just outside the study area and here again some settlement sites are uncertain. Salmonsbury (GR 1720) is known and several others, but whether all camps and all sites where several Dobunni coins have been found, or where traces of early field systems have been recognised, indicate permanent settlement, remains conjectural. Assuming that they do so, nearest neighbour analysis gives an index value of 1·62. This indicates a slightly more regular spacing than the Neolithic (Fig. 22b).

The Iron Age witnessed a succession of invasions, and without datable finds such as pottery or currency it is impossible to tell if sites were used throughout the period. Some camps may have been only temporary shelters for people and livestock. Most sites had good defensive qualities and excellent visibility, and where a sub-optimal site had been chosen it may have provided easier access to water. The numerous bones of sheep and the wool working equipment found in excavations indicate a pastoral economy, while querns, storage jars, ox goads and cultivated fields indicate cereal growing. No significant break is seen between earlier settlements and those of the Iron Age, and sites were in similar localities with a slight bias towards defence or towards possible communication lines. The same tabular format for recording site factors was used as with the other settlement periods.

Roman settlement

Roman settlements were often developed on the sites of the Iron Age, e.g. Wycomb (GR 0220), Salmonsbury (GR 1720) and Norbury (GR 1215), but the villas in the study area were located differently. Villas represent the focal points of estates developed to exploit the cornlands and sheep pastures. Trade in corn and wool took place with or through the local Roman town, in this case Corinium (Cirencester), and road access was, therefore, vital. So villas were located along the Roman roads (Fig. 22c). (Another project has shown that villa numbers peak at 7–10 miles (11–16 km) from Roman towns. This distance represents a maximum convenient transporting distance for produce. Land near to towns may not have been available for private ownership.)

Villas were often located in pairs as Figure 22c shows, reflecting perhaps the same considerations as those discussed by Chisholm (1962) for the Dutch polders with savings on road maintenance costs and increased social contact. With *pax Romana,* defensive sites were no longer necessary and scenic appeal may have become important. Each site had attractive views, at least to modern eyes. All villas were located below springs whence water could flow through the buildings. This was important to the layout of the baths, which characterised the villas. Latin texts suggest that the homes of the estate's work force were normally some distance away from the villa, and the recent discoveries of Romano-British settlements associated with some of these villas may represent these dwellings. A number of villas were located on north east facing slopes and so would have been exposed to the coldest weather.

Nearest neighbour analysis gives a value of 1·21 and with second nearest neighbour a value of 1·92. The difference in value indicates the effect of pairing and the second value shows the very regular distribution of pairs of villas.

Saxon settlement

Saxon villages were identified by their place names, and the date of their earliest documentation was checked from the lists in the relevant volume of *Place-Names of Gloucestershire.* In the five centuries preceding the Domesday Book of 1086, development of the Saxon settlement pattern would have occurred, and as with the other settlement periods the cumulative pattern was studied. Twenty-nine sites were identified (Fig. 22d) and the study area was now packed. Nearest neighbour analysis gives a value of 1·37, which surprisingly does not indicate the most evenly spaced distribution pattern recorded for the four periods.

Again the possible location factors were considered in the field. One striking feature was the coincidence of Saxon villages with the upper boundary of the outcrops of fuller's earth or of Lias clay. The sites were just above the associated springs. Roman roads no longer exerted an influence, and Saxon communities are assumed to have been self supporting and to have obtained life's necessities from their own territories. The criteria for site selection – water, arable land, grazing land, fuel and building materials, discussed by Chisholm (1962) – were checked. A possible change in weighting of these factors on the Cotswolds was suggested because both arable land and sheep pasture are widespread, but meadowland is not. So meadowland would have been a relatively more highly rated factor. It occurs along the clay valley bottoms. Apparently, rather better defensive sites than those of the Roman villas were chosen. With an infilling of settlement, it is likely that later sites may have been sub-optimal because of the need to be sufficiently far from earlier sites to give adequate land space, but place names endings were not as helpful in distinguishing settlement order as they may be in south east England.

Variations in productivity within the different types of land use would have been small, although some exposed parts would have a shorter growing season. So equally sized settlements might be expected to use equally sized territories. The fact that territorial units do not seem to have been equal suggests that the communities were of different sizes.

Later settlement

Northleach (GR 1114) is the only important post Saxon settlement in the area, and was established near a road junction as a thirteenth century market. Eight settlements have been deserted or reduced to single farms, and these were mainly on the more exposed sites, e.g. Roel (GR 075249) and Sevenhampton (GR 024219). Otherwise the Saxon settlement pattern remains today.

It was concluded that the settlement patterns of the four periods differ, and at each period the particular economy, technology, population numbers and vulnerability led to a slightly different evaluation of the land's resources. Human decision making in site selection is too complex to be replicated exactly but the major constraints on the choice of a site can be inferred from the resulting settlement pattern.

Comment

In evaluating this study several comments should be made. It took more time than one could normally be expected to give to a project and most of the site factors of settlements could have been identified from the map. However, the 'feel' of a site can only be obtained in the field where the full scale of the environment can be seen. Many rejected hypotheses in archaeology have originated from work on small scale maps where distance has not been fully appreciated.

In dealing with archaeological distribution patterns, one cannot be certain of the extent of monument destruction that has occurred, and new discoveries may radically alter the patterns. So inferences drawn from them must be tentative.

The settlement types studied here are not exactly comparable. For example, a villa is not a village, and there were village type settlements in the Roman period. Nor were long barrows villages, and some Iron Age camps may not have been lived in. But it could be argued that the features considered here would have been the principal settlement elements seen by each new wave of colonists. The sixth-former had little alternative than to use these.

The original expectation was that nearest neighbour analysis would show a gradual development from the most irregular distributions in the Neolithic and Iron Ages, to regularly spaced settlements in the Saxon period. The sequence of R_n values of 1·48, 1·62, 1·21 and 1·37 does not show this. One must add, however, that the technique is not entirely satisfactory for this purpose, the chief difficulty being a variation of the density of sites at the different periods.

Study 5 A test of Christaller's central place theory?

Human geography text books normally contain a brief review of central place theory and an explanation of Christaller's basic concepts of central place, complementary regions, range of a good, and economic distance. They also contain diagrams of the settlement patterns based on the three principles of marketing, traffic and separation (administration). The theory is introduced as a means of identifying order in the distribution patterns of settlements of various sizes and degrees of importance, and its simple logic attracts interest. But does it work? This is the question one is prompted to ask, and in this study a sixth-former has attempted to apply the theory to a particular region to see if the theoretical settlement pattern fits the actual pattern. As a project it may be regarded as typical of those that seek to test a well-known theory in a local context.

Some details of the theory and of its assumptions

Summaries of Christaller's work did not give sufficient practical details of how to apply the theory to an area, so the initial task was to obtain a copy of the translation of the original work and to read it carefully. This took several weeks to complete but was judged to have been worthwhile as, in addition to giving the required details, the book also demonstrated the logical development of a research programme, and showed Christaller's critical awareness of weaknesses in the theory. An impression of thorough, painstaking, honest and clear-thinking research was gained by reading the original work.

The theory made assumptions of an environment where physical conditions and resources were uniformly distributed and where the population was settled in discrete units farming the surrounding land. Initially, these units were to be evenly distributed, self contained and of equal population, but later commercial enterprise by one or more of the settlements would trigger off the development of central places. Christaller was aware that many factors were involved in settlement distribution, but the above assumptions were made in order to emphasise the role of economic factors. Such an area where terrain and agricultural resources are more or less uniform and where settlements are nucleated and regularly spaced (nearest neighbour index of 1·83) is Beauce, a *pays* south west of Paris. This seemed to be an ideal area in which to test central place theory and a field investigation was, therefore, planned. This took place in the summer vacation 1975.

Central place theory is deductive in that it is concerned with the application of economic principles to the spacing of settlements. Different goods and services have different ranges, i.e. the economic distance that people are prepared to travel for the goods will vary with the type of goods. For example, people will be willing to travel further for furniture than for groceries. Also the numbers of people necessary to support shops selling particular types of goods will vary. The threshold population (the minimum population to make a shop economically viable) for a furniture store will be much higher than that for a grocers. Christaller deduced from these two concepts that settlements containing high order goods would serve a larger area and a larger population than settlements containing only low order goods, and he demonstrated that a high order central place would serve a large complementary region within which will be low order central places surrounded by their own much smaller complementary regions. He envisaged a hierarchy of central places with their regions fitting together as in Figure 23.

To measure the centrality of a central place Christaller used the number of telephone links between it and the surrounding region. He argued that centrality is the measure of importance of the central place to its surroundings and that telephone connections would indicate centrality more clearly than would population totals or numbers of functions. Few people in southern Germany, where the research was undertaken, possessed telephones in 1933, and those that did used them primarily for business purposes. The actual number of telephones in a central place did not give the measure of centrality, but rather the excess number of telephones over the mean for the population. This was calculated from the formula

$$Z_z = T_z - E_z \, (T_g/E_g),$$

where Z_z is the centrality index, T_z is the number of telephones at the central place, E_z is the number of inhabitants of the central place, T_g is the number of telephones in the complementary region, and E_g is the number of inhabitants of the complementary region.

By calculating centrality indices, Christaller was able to classify settlements as central places of different orders ranging from the lowest M- (market town), through A-, K-, B-, G-, P-places to the highest R-place (world city). He was also able to calculate the average distances between central places of each order beginning with M-places which were 7 km apart – this was based on a complementary region of 4 km radius which corresponded to one hour's travel on foot. Radii increased by $\sqrt{3}$ km for each higher order central place in the

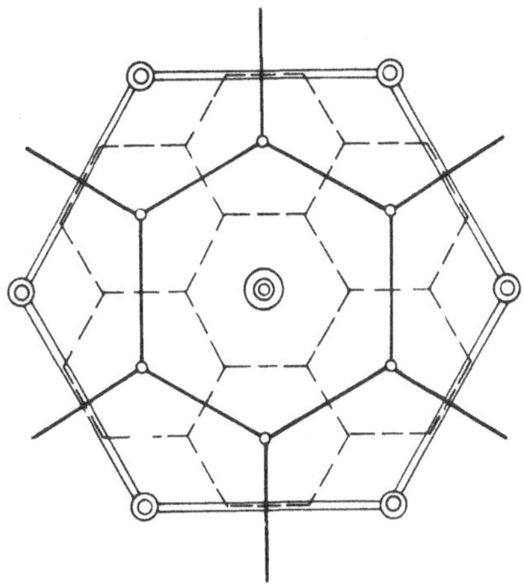

Figure 23 Central place hierarchy and associated market areas.

hierarchy, giving a series of 4, 7, 12, 21, 36, 62, 108 and 185 km. The complementary regions, in order to completely serve land areas without overlapping were hexagonal in shape.

Application to the Paris Basin

With this basic information, it was now possible to begin to apply the theory to Beauce. Paris was the R-place and, therefore, the theoretical settlement pattern was drawn around Paris. The Paris Basin as an agricultural region developed around Paris as the market, and one would, therefore, expect a marketing principle to have operated from early days. Six lower order central places (P-places) would be expected at 108 km from Paris. In fact, Rouen is 115 km distant and Amiens, Reims, Troyes and Orleans are 114, 129, 136 and 109 km distant respectively. There is no P-place west south west of Paris, but Chartres, a G-place, lies between the theoretical locations of G- and P-places and probably fulfils some of the functions of the missing P-place (Fig. 24). The area of field work was to be south east of Chartres, and a theoretical central place system was, therefore, drawn around Chartres as the G-place. A ring of B-places 36 km from Chartres should exist, and Rambouillet at 34 km, Dreux at 34 km, La Loupe at 35 km and Bonneval at 30 km represent this ring. Similarly a ring of K-places would be expected at 21 km distance, and in the study area Voves is 21 km from Chartres. Again, around Voves as the K-place, a theoretical settlement pattern was drawn to include A-, M- and H- (village) order central places as in Figure 25.

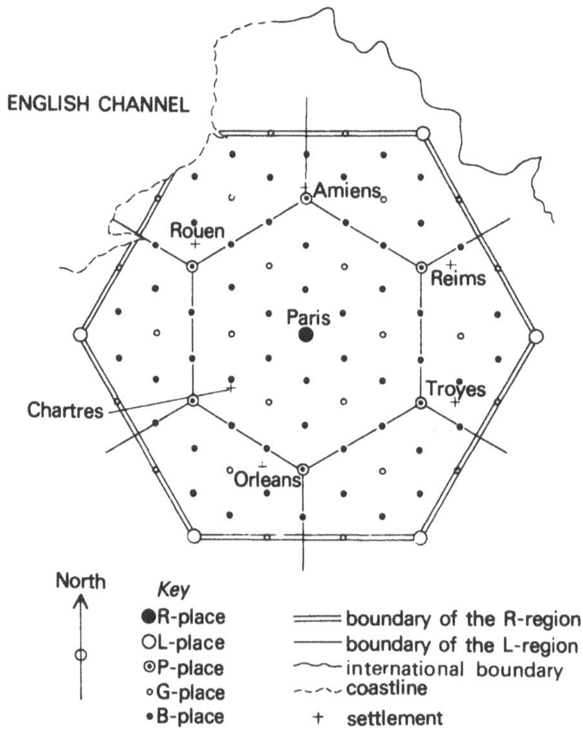

Figure 24 Central places in the Paris Basin.

This map was then placed as an overlay on to the Carte Touristique 1:100 000 Paris–Orleans sheet to see if the distribution of the theoretical and actual settlements coincided. None did precisely but at this scale many were very close indeed.

Data collection

To progress from this stage, one required information on the centrality of the settlements in the Voves area. Two methods of investigation were possible. First, each settlement could be visited and the number of central functions recorded from observations there, and secondly, a calculation based on Christaller's formula could be made using data in the Eure-et-Loir telephone directory. Both methods were used.

With help from other sixth-formers, lists of services and functions were compiled for each of the seventeen central places in the study area. A small sample of people in each settlement was asked where groceries, clothing and furniture were normally purchased – each

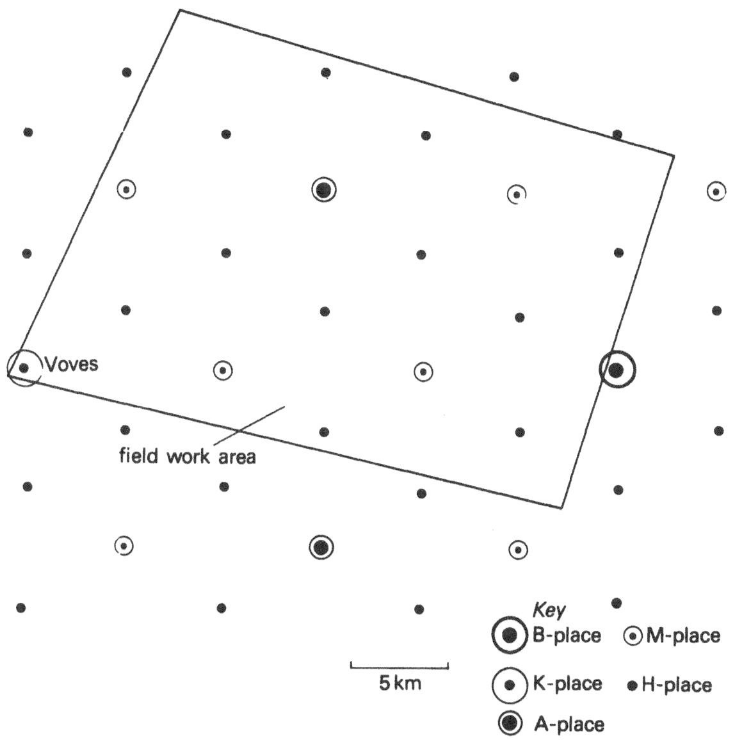

Figure 25 Expected distribution pattern of central places near Voves.

product being an indicator of short, medium and long range goods respectively. The locations of the nearest secondary school, hospital and theatre were also found by inquiry. From this survey the number of central functions at each settlement was totalled as follows:

Baudreville	13	Moutiers	4
Beauvilliers	4	Neuvy	2
Boisville	13	Ouarville	16
Fresnay L'Evêque	13	Prasville	7
Gouillons	5	Réclainville	1
Levesville	2	Trancrainville	2
Louville	3	Voves	61
Mérouville	0	Ymonville	16
Moinville	2		

A hierarchy of central places was now clear with Voves as the K-place, Ymonville, Ouarville, Fresnay, Boisville and Baudreville as M-places, and the remainder as H-places of small centrality. When these were plotted on a map using the same symbols as in the theoretical map, a more impressive correspondence of theory and reality was shown (Fig. 26). No attempt was made to measure the degree of correspondence though a method might have been devised.

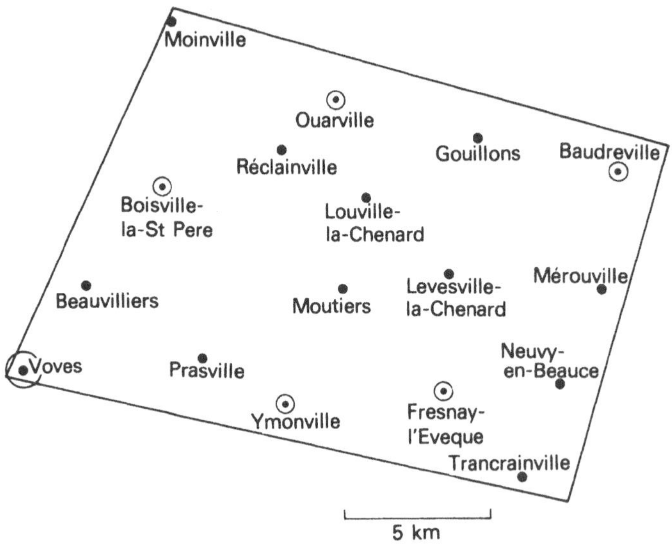

Figure 26 Actual distribution pattern of central places near Voves.

Calculation of centrality indices

The departmental telephone directory, in addition to giving the names, addresses and numbers of subscribers, also records the nature of businesses. It is, therefore, possible to scan the directory for the number of central functions in each settlement. Christaller's formula was then applied to the survey area using the numbers of telephones with business or service functions in each settlement and population numbers obtained from the latest census. Centrality indices were then calculated as shown in the table.

Several problems arose during this calculation and some of these are inherent to the whole theory. First, the boundaries of the complementary regions, which for practical reasons were taken to correspond to the administrative boundaries of the communes, may not have been accurate. The centrality index depends on the choice of the regional boundary and the population sample was not large enough to determine this effectively. Secondly, the proportion of the commune's population living in the central place to that of the region could only be estimated for the lowest order central places, although with nucleated settlements this was relatively easy to do. For higher order central places this problem diminished but the regional boundary difficulty remained. Another problem with low order central places occurred when the central place contained all the business telephones of the

Settlement	*Pop. of settlement*	*No. of business telephones in settlement*	*Pop. of region*	*No. of business telephones in region*	*Centrality index*
Baudreville	150	15	200	16	3·0
Beauvilliers	180	5	300	9	– 0·4
Boisville	420	18	600	23	1·9
Fresnay l'Evêque	470	17	600	19	2·1
Gouillons	200	5	200	5	0·0
Levesville	160	4	200	5	0·0
Louville	180	5	200	5	0·5
Mérouville	190	5	200	5	0·25
Moinville	100	2	100	2	0·0
Moutiers	120	4	200	6	0·4
Neuvy	280	4	300	4	0·3
Ouarville	300	25	400	27	4·75
Prasville	290	8	300	8	0·3
Réclainville	90	4	100	4	0·4
Trancrainville	280	3	300	3	0·2
Voves	2500	111	10000	375	17·25
Ymonville	550	24	600	24	2·0

community, and this gave a higher centrality index than for some settlements where, although many more central functions were provided by the central place, some were also found in the surrounding area. Again, with higher order central places the difficulty disappeared, and it was noted that central place theory was developed with the higher order settlements chiefly in mind.

Conclusion

Generally the centrality index gave a similar hierarchy of settlements to the field observations, and when the distribution patterns of the theoretical central places and the actual were compared, it was found that the M-places Fresnay and Boisville were within a kilometre of their predicted sites, and Ymonville and Baudreville only two or three kilometres distant. Ouarville, the other M-place according to observations, was two kilometres from a theoretical A-place, and it was noted that its business telephones were more numerous than those for the other M-places – perhaps it was an A-place. So some visible correspondence between the theoretical and the observed settlement patterns could be seen within the study area. The project was concluded with a discussion on the possible causes of differences between the two patterns. A decline in the agricultural population and a rise in the numbers of Paris commuters living in some of these villages would have meant improved economic means and purchasing power in recent years, so economic distances would have changed.

Increased mobility through car ownership would have greatly extended the complementary regions of the higher order central places. The supermarket in Chartres would have displaced the village shop. During the field work several mobile shops were noticed where the goods were being brought to the consumer rather than the consumer travelling to the central place to buy them. This too may have distorted the patterns of central functions in lower order places. It has been suggested that the settlement pattern of Beauce was derived from the pattern of Gallo-Roman estates on which corn was grown for distant markets. If this is so, the estates would have been linked to the Roman road network and Christaller's traffic principle may have been in operation at the time of settlement origins.

Perhaps the pattern of observed central places should have been compared with the pattern for Christaller's traffic principle rather than with the pattern for the marketing principle. Projects often turn up a new problem just when the end is in sight.

Study 6 The distribution pattern of monastic estates

On reading Atkyns' *Ancient and Present State of Glostershire* 1712, one cannot fail to notice the references in parish after parish to lands which were formerly held by monastic houses. Some were held by distant owners such as the Archbishop of York, or the Abbot of Westminster, others belonged to overseas monasteries such as St Denis in Paris, and Fecamp in Normandy, but the majority were possessions of the local monastic houses such as St Peter's (Benedictine), now Gloucester Cathedral, Winchcombe (Benedictine), Tewkesbury (Benedictine), Deerhurst (Benedictine), St Oswalds and Llanthony (both at Gloucester and both Augustinian), and Hailes (Cistercian). The project considered in this study investigates the spatial pattern of the different estates and tackles the question: *Do geographical factors explain the distribution of monastic estates in north Gloucestershire?* It may be viewed as a typical historical geography project which seeks to examine spatially the data contained in the written record of an early survey. Many similar projects can be envisaged.

There was an unusual concentration of monastic houses in Gloucester and the neighbouring area during the medieval period, and using the information contained on the OS sheet Monastic Britain, a chi square test was applied to ascertain whether or not this concentration was statistically significant. The north Gloucestershire area, selected for reasons of local interest, was taken as the central square in a grid of nine, and with 17 of the 56 monastic houses in the total area, a chi square reading of 27·05 was obtained – this is significant at the 0·01 level. The null hypothesis was, therefore, rejected and this concentration of monastic houses shown to be statistically significant. The reasons for the clustering were not considered, but clearly, with such a concentration of monasteries, many monastic estates would be expected to occur within the study area, and the project now turned to the expected patterns of distribution of these estates.

Factors expected to influence the distribution of estates

Monastic estates were acquired at the time of the initial foundation charter and from subsequent donations. Rarely were they purchased, although there is some evidence, for example at Hailes, of appeals for estate appropriation in times of economic difficulty. One might expect that the greatest gifts of land or the greatest number of gifts would come from the areas of greatest prosperity. One might also expect that donations would come from nearby rather than from afar where the influence of the monastery would have been less conspicuous, and that the quality of land would have some bearing on the distribution of bequests. Older foundations would be expected to have received a larger number of gifts of land, and also to have had a less scattered arrangement of properties than the later foundations when land available for donation would have been more limited. With the growth of the overseas wool trade in the thirteenth century, one would expect that monastic estates would expand across the Cotswolds, the great wool producing area, and also along the major trade routes. These expectations were set out as hypotheses to be tested once the distribution of monastic land holdings had been determined.

The distribution of estates

Several sources were consulted for the information on holdings including the *Cartularies of Gloucester Cathedral, Valor Ecclesiasticus,* and Dugdale's *Monasticon Anglicanum* – available either in the county records office or the cathedral library. In none of these references was a complete list of properties found for all of the monasteries and, in the end, it was decided to base the distribution map of estates on the details contained in Atkyns' book. It was not possible to verify the accuracy of these details but since Atkyns' sources included the notes of Richard Parsons, Chancellor of the Diocese of Gloucester, they are likely to be reliable. The style of presentation is rather imprecise as this entry for Prestbury shows: 'Divers lands in Presbury, and the Impropriation, did formerly belong to the priory of Lanthony, who had Free Warren in the Lands belonging to the Priory 15 E1. It was appropriated to that monastery 21 R2.' The exact locations of the estates and their areas are not given by Atkyns so in mapping, a different symbol was used to represent each of the monasteries and these symbols were placed at the centre of each parish where the monastery had land. Figure 27 is an extract from the map produced.

Hypothesis testing

With this map as a basis, work on hypothesis testing could begin. The distribution of estates in relation to distance from the monastic house was examined first. Concentric rings at 2 mile (3·2 km) intervals were drawn around each monastery, the areas of each circular band were calculated, and then the density of estates within each band was

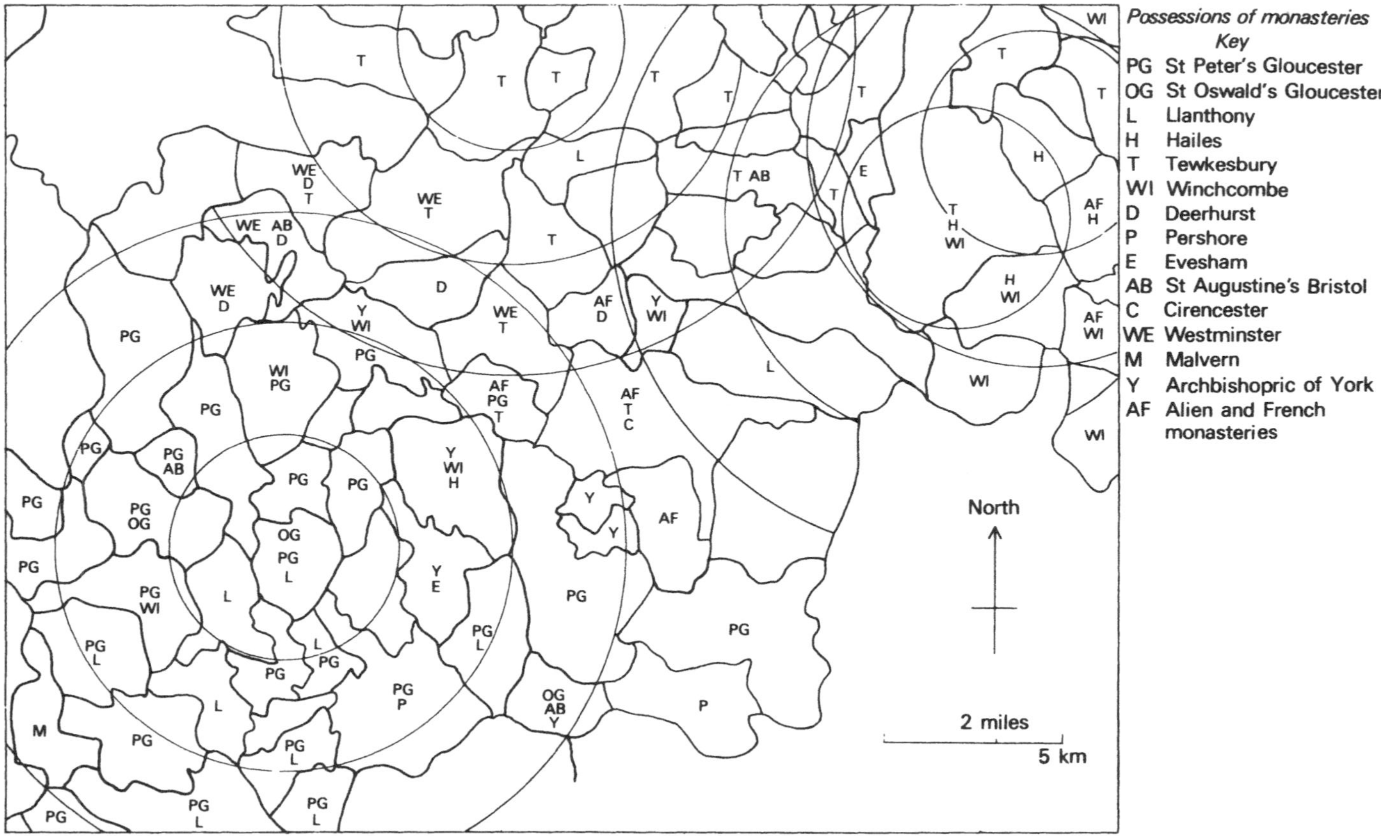

Figure 27 Distribution of monastic estates in part of north Gloucestershire (based on Atkyns 1712).

plotted on a histogram. The estates of all the monasteries show clear distance decay trends (Fig. 28). It would have been possible to show on the same graph the trend lines of the relationship between estate density and distance from monastery for all the monasteries, and thus to see if the earlier foundations had a more compact distribution of estates than the later ones by their predicted steeper trend line gradients. This was not done but inspection of the separate histograms suggested that this was not the case.

The second hypothesis tested was that the donations came primarily from the areas of greater prosperity. Here, another data problem was encountered in the search for evidence of prosperity. It was argued that the richer agricultural areas would have had a larger proportion of land in arable use and that these might, therefore, be identified as areas with a higher density of plough teams per village as recorded in the Domesday Book (1086). Darby's analysis of the Gloucestershire Domesday Book contains a map of the plough team density in the county, and this map, together with the details contained in Domesday, were used as the data source for agricultural prosperity.

The parishes of the study area were divided into three groups according to their density of plough teams, and the percentage of the study area falling into each group was then calculated. Then, the total number of monastic estates (127) in the area was divided proportionally among these three groups. This gave the expected number of monastic estates in each group, assuming that they were evenly

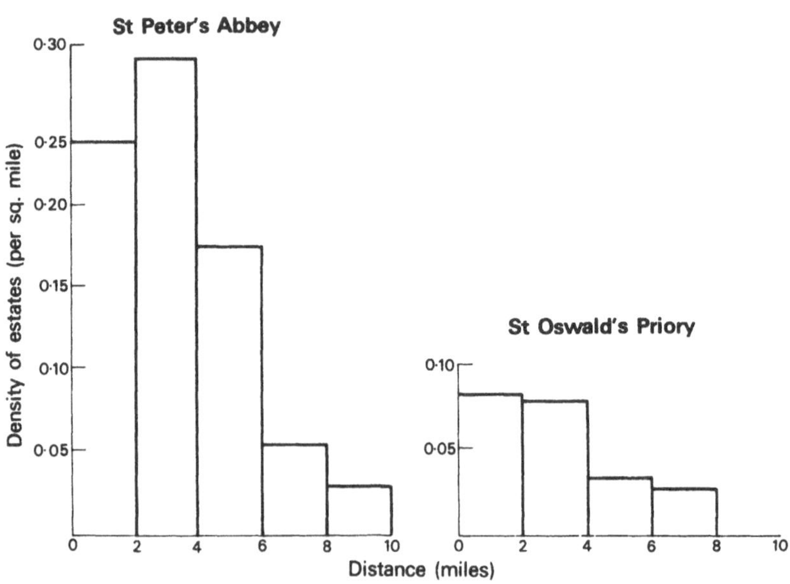

Figure 28 Distance decay of monastic estates.

distributed. Now a chi square test was applied comparing the actual distribution with the expected distribution. The calculation is as follows:

Plough teams per sq. mile	*Observed frequency*	*Expected frequency*	$O-E$	$O-E^2$	$\frac{(O-E)^2}{E}$
Below 2·5	73	63	10	100	1·59
2·5 – 5·0	39	48	−9	81	1·69
Above 5·0	15	16	−1	1	0·06

$$\chi^2 = \sum \frac{(O-E)^2}{E} = 3{\cdot}34$$

A chi square reading of 3·34 was not high enough to be statistically significant at the 0·05 level, and so the null hypothesis, that there is no relationship between the distribution of monastic estates and the distribution of arable land (used here as an index of prosperity), could not be rejected. The general fertility of land did not seem to directly influence donations of estates.

Meadowland was frequently the most highly valued type of agricultural land because of its relative scarcity in many areas. No details were available of the amount of meadowland in the monastic holdings nor of the quantity of meadowland in those parishes from which land was donated to the monastic orders. Atkyns did, however, record whether or not meadowland occurred in his day in the various parishes, and these details are likely to be relevant for the medieval period also. *Valor Ecclesiasticus* gave the values of estates held by monasteries in each parish at the time of the dissolution. Given these values and the dichotomised data of whether a parish had meadowland or not, it was possible to measure the correlation between estate values and meadowland. The point-biserial correlation method was used. If a significant relationship had been found it could have been interpreted in two ways, either that meadowland was so valuable that any estate containing meadows would be highly valued, or that estates containing meadowland were larger and so more valuable. In fact, the correlation coefficient was −0·03 so there was no significant relationship between the distribution of monastic estates and of meadowland.

To plot the diffusion of monastic estates, and so to test whether or not the wool producing land of the Cotswolds figured as prominently in later bequests as might have been expected from the records of the large flocks of sheep held by the monasteries, required information on the dates of land acquisition. Insufficient data were found to test this idea. Similarly, the hypothesis that estates expanded along the major routeways, for which there is some secondary evidence, could not be tested without the dates of estate acquisition.

A land quality classification of the county has been made, and a map has been published in the *Gloucestershire Survey* by G. E. Payne to show the distribution pattern of land quality. This was inspected for relationships with former monastic estate distributions by the use of overlays but there was no clear visible relationship.

Conclusion

The project concluded therefore, that the only geographical factor that could be shown to have a bearing on monastic estate distributions was distance from the monastery; that the distributions of indices for land value, land use and quality of soil did not correlate significantly with monastic lands; and that data shortage prevented the construction of an estate diffusion model. It was noted that the apparent precision of the statistical analysis should not mask the fact that the data used may not always be taken at face value – a competent medieval historian would have identified the uncertainties.

Typically, new ideas for investigation were uncovered in addition to the obvious need for a more exhaustive search for estate acquisition dates. It was observed that the Benedictine order obtained the ownership of manors and so received its agricultural wealth from feudal dues – the Cartulary of St Peter's, for example, recorded the rights over tenants exercised by the Abbey. On the other hand, the Cistercian order, for example at Hailes, farmed directly from granges. The agricultural influence of the different orders varied, therefore, and these variations would have been expressed spatially.

The distribution of monastic lands among new land owners as a result of the dissolution would also offer scope as a project topic.

Records of sheep movement between summer pasture on the Cotswolds and winter pasture in the Vale were also found and so a new type of estate linkage was suggested. This would have affected the development of medieval routeways and many of these former sheep tracks will be preserved in the present day road and path network.

Finally, it was suggested that the distribution of lands of other large landowners of different historical periods such as some Oxford and Cambridge Colleges, financial institutions, and wealthy families might be studied in a similar way.

Study 7 An analysis of the changes in connectivity of a road network

The ideas and techniques used in this project, like those included in Study 5, come straight from the normal work of an A-level course. The syllabus required an understanding of transport networks, of their development and of their effects on the economy of the areas served.

Some transport systems, such as Britain's railways in the early 1960s and rural bus services in the 1970s, have declined, and offer scope for projects measuring the reduction in accessibility by public transport. On the other hand, motorway developments and new bridge constructions offer scope for monitoring the effects of increased accessibility. In the summer of 1971 an opportunity arose of spending one day of a field course in the Delta region of the Netherlands. Here, more conspicuously than in any other part of Europe, one might expect to see the effects of improvements in a communication system.

The islands of Zeeland used to be connected to each other and to the mainland by ferry crossings, but with the implementation of the Delta Scheme a network of modern roads crossing new bridges and dams has been built to improve these connections. With the completion of each road link accessibility has improved, and using the detailed information provided by the Netherlands Government Information Service the stages in the road network development can be reconstructed easily (Fig. 29).

Expected effects of increased accessibility

One might expect that a new road link to an area, which was an island until the second half of the twentieth century would have a marked effect on the social and economic geography of the area and especially when the former island is as near to a major urban centre as the northern islands of Zeeland are to Rotterdam. People working in Rotterdam could commute with ease from such an island, and its population would, therefore, increase and new house building programmes would commence to meet the needs of the enlarged population. The prices of older property would be expected to rise and this would be associated with cottage renovation. The age structure of the island would also change with the arrival of younger families. Improved access would encourage industry because materials and

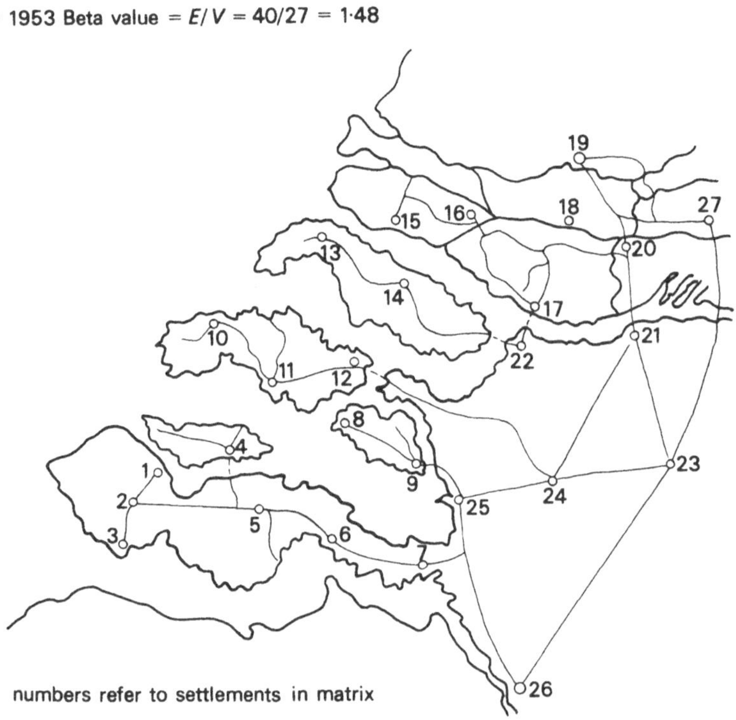

Figure 29 (a) Original road network; (b) Delta scheme road network 1971.

products could now be transported more conveniently by lorry, and agriculture would benefit in a similar way as the rapid transport of market garden produce to urban markets would now be possible. One might also expect a rise in the recreational interest shown by the major population centres in these nearby, but formerly isolated, parts. Social problems associated with these economic changes and with the rapidly altered social structure would need attention from the government and one might expect to find evidence of such care.

Simple ideas like these formed the basis of the project and were fashioned into statements of expectation that could be treated as hypotheses to be tested against field observations. Field work opportunities were of necessity very limited and the sample locations, where hypothesis testing was to take place, would be few in number. A wise choice of these locations could only be made if the spatial variation of improvement in accessibility over the Delta area was known, and this, therefore, became the first item for investigation in the project.

(b)

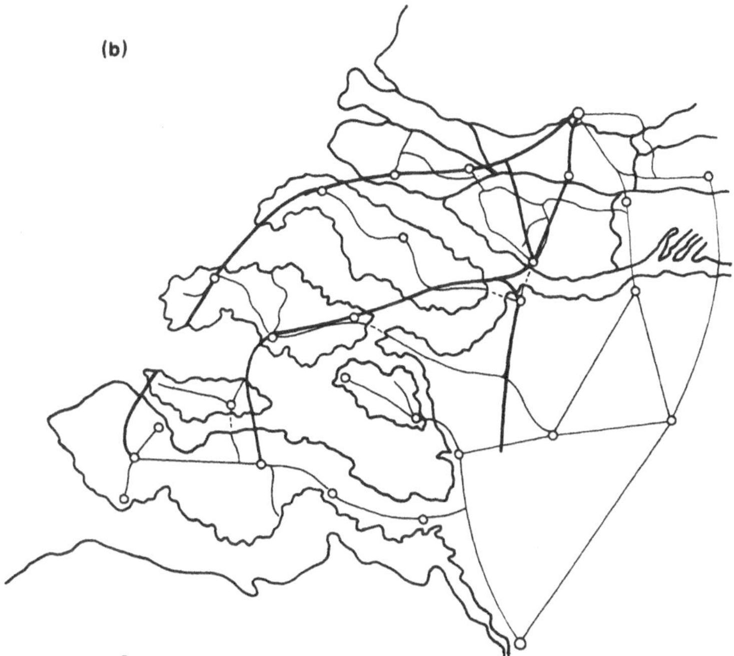

1971 Delta Scheme Road Network
original road network

1971 Beta value = $E/V = 53/27 = 1{\cdot}96$

Mapping accessibility

Measurement of improved accessibility implied measurement, first, of the original communication network and then, secondly, of the final network – the measurement of improvement being the difference between the two. The technique of network analysis was chosen for this task. In network analysis each settlement is referred to as a node, or vertice, and each road connection between two adjacent settlements is a link, or edge. As connectivity is improved the number of edges increases while the number of vertices may remain the same. Thus the Beta index, which is obtained from the number of edges divided by the number of vertices (E/V), for the Delta area, increased from 40/27 or 1·48 in 1953, to 53/27 or 1·96 in 1971. Connectivity for the whole area increased, therefore, by 1·3 times. But more important than the overall change was the varying degree of change within the area.

To determine this the connectivity of each settlement had to be measured both before and after the scheme. A topological map was produced for the Delta which used as vertices the main villages and towns of the islands and the towns of the adjoining mainland. Then the number of edges that were traversed to reach every other settlement from each individual settlement was counted, and recorded in a matrix with the totals noted in the last column as shown in the table (Fig. 30).

	1	2	3	4	5	6	7	8	9	10	11	12	13	14	15	16	17	18	19	20	21	22	23	24	25	26	27	*Total* 1971	*Total* 1953	*Difference*
1. Veere	0	1	2	2	2	3	4	7	8	4	3	4	5	5	6	6	5	6	7	6	7	5	6	5	5	5	7	130	184	54
2. Middelburg	1	0	1	1	1	2	3	6	7	3	2	3	4	4	5	5	4	5	6	5	6	4	5	4	4	4	6	103	145	42
3. Vlissingen	2	1	0	2	2	3	4	7	8	4	3	4	5	5	6	6	5	6	7	6	7	5	6	5	5	5	7	129	169	40
4. Kortgene	2	1	2	0	1	2	3	6	7	2	1	2	3	3	4	4	3	4	5	4	5	3	4	3	4	4	5	91	173	82
5. Goes	2	1	2	1	0	1	2	5	6	2	1	2	3	3	4	4	3	4	5	4	5	3	5	4	3	3	6	87	124	37
6. Hansweert	3	2	3	2	1	0	1	4	5	3	2	3	4	4	5	5	4	5	6	5	4	3	4	3	2	2	5	93	109	16
7. Bath	4	3	4	3	2	1	0	3	4	4	3	3	4	3	5	4	3	4	5	4	3	2	3	2	1	1	4	84	95	11
8. Stavenisse	8	7	8	7	6	5	4	0	1	7	6	5	6	5	7	6	5	6	7	6	5	4	5	4	3	4	6	143	152	9
9. Tholen	7	6	7	6	5	4	3	1	0	6	5	4	5	4	6	5	4	5	6	5	4	3	4	3	2	3	5	118	128	10
10. Renesse	4	3	4	2	3	4	5	6	7	0	1	2	1	2	2	3	3	4	4	4	5	3	5	4	4	5	5	95	161	66
11. Zierikzee	3	2	3	1	2	3	4	5	6	1	0	1	2	2	3	3	2	3	4	3	4	2	4	3	3	4	5	78	137	59
12. Bruinisse	4	3	4	2	3	4	3	4	5	2	1	0	2	1	3	2	1	2	3	2	3	1	3	2	2	3	4	69	114	45
13. Goedeveede	5	4	5	3	3	4	4	5	6	1	2	2	0	1	1	2	2	3	3	3	4	2	4	3	3	4	4	83	159	76
14. Middelharnis	5	4	5	3	3	4	3	4	5	2	2	1	1	0	2	2	1	2	3	2	3	1	3	2	2	3	4	72	135	63
15. Hellevoetshuis	6	5	6	4	4	5	5	6	7	2	3	3	1	2	0	1	2	3	2	2	3	3	4	4	4	5	3	95	156	61
16. Spijkenisse	6	5	6	4	4	5	4	5	6	3	3	2	2	2	1	0	1	2	1	1	2	2	3	3	3	4	2	82	132	50
17. Numansdorp	5	4	5	3	3	4	3	4	5	3	2	1	2	1	2	1	0	1	2	1	2	1	3	2	2	3	3	68	123	55
18. Barendvecht	6	5	6	4	4	5	4	5	6	4	3	2	3	2	3	2	1	0	1	2	3	4	4	3	3	4	2	91	—	—
19. Rotterdam	7	6	7	5	5	6	5	6	7	4	4	3	3	3	2	1	2	1	0	1	2	3	3	3	4	4	1	99	128	29
20. Dordrecht	7	5	7	4	4	5	4	5	6	5	3	2	3	2	2	1	1	2	1	0	1	2	2	2	3	3	1	82	99	17
21. Moerdijk	7	6	7	6	5	4	3	4	5	5	4	3	4	3	3	2	2	3	2	1	0	2	1	1	2	2	2	89	91	2
22. Willemstad	6	4	5	3	3	3	2	3	4	3	2	1	2	1	3	2	1	4	3	2	2	0	2	1	1	2	3	68	90	22
23. Breda	7	6	7	5	5	4	3	4	5	5	4	3	4	3	4	3	3	4	3	2	1	2	0	1	2	1	1	92	93	1
24. Roosendaal	6	5	6	4	4	3	2	3	4	4	3	2	3	2	4	3	2	3	3	2	1	1	1	0	1	2	2	76	78	2
25. Bergen op Zoom	5	4	5	4	3	2	1	2	3	4	3	2	3	2	4	3	2	3	4	3	2	1	2	1	0	1	3	72	82	10
26. Antwerp	5	4	5	4	3	2	1	3	4	5	4	3	4	3	5	4	3	4	4	3	2	2	1	2	1	0	2	83	95	12
27. Gorinchem	7	6	7	6	6	5	4	5	6	5	5	4	4	4	3	2	3	2	1	1	2	3	1	2	3	2	0	99	111	12

Figure 30 Connectivity matrix of Delta region 1971.

Edges representing ferry crossings were arbitrarily regarded as double. The figure shows the connectivity totals for 1953 and 1971, and by subtracting the former from the latter, the changes in connectivity for each settlement as a result of road development can be seen. For example, Veere shows an improvement of 54 with readings of 184 and 130, and Zierikzee an improvement of 59 with readings of 137 and 78. Maps showing old and new connectivities and and the degree of improvement in connectivity were produced using isopleths of equal connectivity totals as in Figure 31. With the delineation of areas of maximum connections and of maximum improvement in connectivity, it was now possible to select the settlements for field survey.

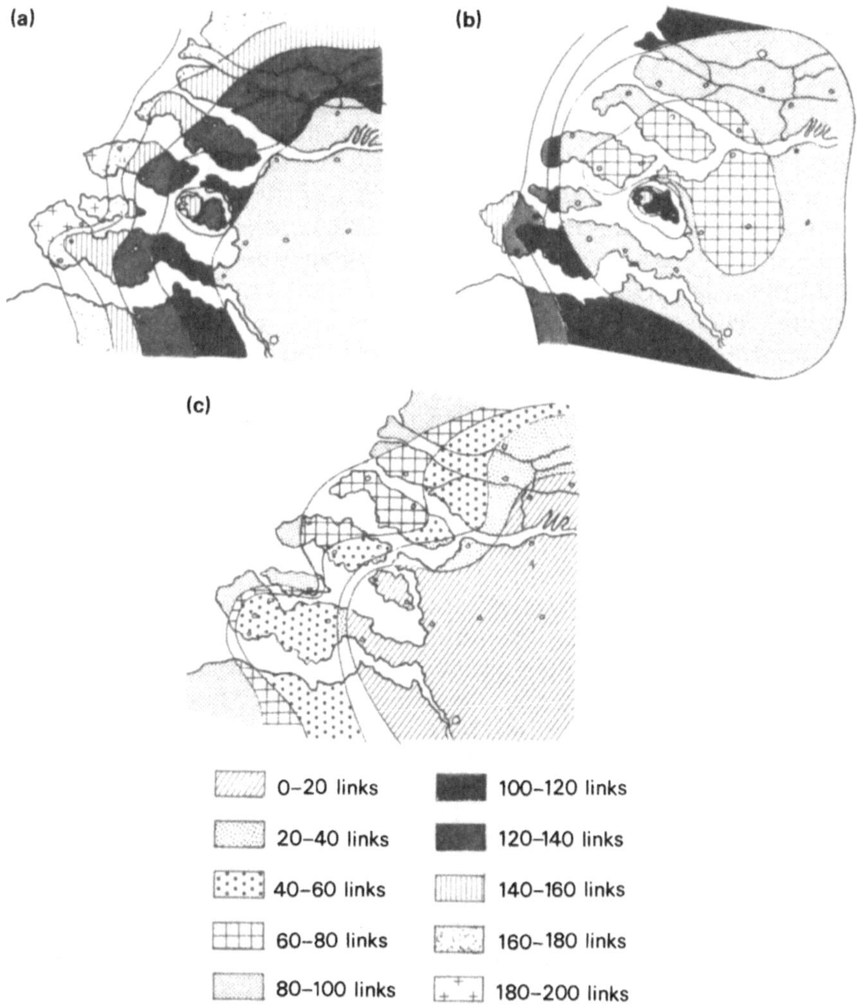

Figure 31 (a) Connectivity of Delta region 1953; (b) connectivity of Delta region 1971; (c) improvement in connectivity between 1953 and 1971.

Field observation

Careful organisation of data collection was essential. Six villages were chosen for sampling on the Island of Overflakkee where connectivity improvements ranged from 20–80 edges, and where connectivity totals were mostly between 60 and 80, i.e. where the effects of improved accessibility were likely to be most conspicuous. Two sixth-formers were allocated to each village with instructions to collect information on the following items: the number of houses in the village and the number built since 1953; prices of property for sale; distances to a selection of amenities including doctor, bank, secondary school; employment structure of the village; and evidence of social activities. They were also asked to talk to as many villagers as possible to gain impressions of the social and economic changes experienced since the bridges had been completed.

Information based on the groups field observation was reliable, but the quality of the impressions varied with the sensitivity of the sixth-formers and with the standard of English of the interviewees. As is normally the case, the pair which included the project leader gained far more than the others. In five of the six villages between a half and a third of the houses had been built since 1953. Prices for property were normal when compared with similar houses in other parts of the Netherlands visited on the field course, but had shown marked increases from formerly low levels. Many young families were recorded. Most men commuted to Rotterdam for work. The data seemed thus to confirm the original expectations.

Evaluation

Further work on the effects of improved accessibility was undertaken after the field visit using the published information of government organisations. For example, graphs of population totals against time were plotted for sample areas within the Delta. These showed that growth had occurred since the Delta Scheme began and that the earlier declining trend had been reversed, but that the areas of greatest increase in population did not correspond to areas of greatest nor most improved accessibility. Published statistics of traffic density for sample areas of the Delta, again, did not correspond to predictions based on the connectivity maps. A check on industrial development showed that this was concentrated in the extreme south west of the Delta on Walcheren where connectivity was poorest, and the inference was drawn that more is involved in industrial development than road improvements.

These latter findings were used to evaluate critically the original hypotheses. In particular, it was noted that the connectivity indices did not include details of the quality of the roads nor of the distances involved; that water transport may still have an important role; and that the effects of improved accessibility could not be isolated from the general development of the Dutch economy nor from the other benefits of the Delta Scheme such as flood prevention, soil improvement and improved navigation. Nor could the break up of the closely knit island communities be entirely attributed to greater road accessibility – less isolated areas have experienced similar changes.

The exercise of measuring connectivity changes was worthwhile and the initial field observations indicated the direction of economic and social change. Perhaps another field course will reap the benefits of this project.

Glossary

aliquot The measured solution placed in a conical flask before titration. Titre refers to the amount of solution added from the burette.
comparison goods Goods which have been purchased after they have been compared with other brands and prices.
competence The weight of the largest fragment of rock a river can transport.
confidence level A level of probability; e.g. a 95% confidence level means that there is only a 5% probability that the result could have occurred by chance.
connectivity The degree to which the nodes or vertices of a network are directly connected.
contingency table A matrix designed to show whether two criteria are independent of one another.
co-ordinates A pair of lines drawn at right angles to each other from the base and side axes of a map by which a point is located. With weighted co-ordinates the lengths of the co-ordinates are multiplied by a factor indicating the importance of the point.
covariance Where two sets of factors vary together.
deductive Reasoning which proceeds from the general to the particular. Inductive reasoning draws generalisations from particular instances.
degrees of freedom The number of variates which must be given values before the complete set of variates can be determined.
EMF Electromotive force.
histogram A bar graph which shows the frequency of items in each class.
hypothesis An idea which can be tested to establish its truth.
interpolate To insert values between measured values. To extrapolate means to calculate values outside the range of those which have been measured.
interval scale A scale on which the exact spacing of data can be measured.
isopleth A line drawn through points on a map which have been calculated to have equal values.
matrix A table of numbers arranged in columns and rows.
normal distribution A distribution of data in which the mean, median and mode coincide, and in which 68% of the data lies within ± 1 standard deviation of the mean. The normal distribution curve is shaped like Napoleon's hat.
parametric test A statistical test which may be applied only to normally distributed data.
regression line The best fit line drawn through the scatter of points on a graph. It expresses the relationship between the two variables.
SMD Standard Man Days. The number of 8-hour working days required for a crop per hectare or for each head of livestock during one year.
superordinate A higher order in classification.
surrogate A more generalised substitute for the required data.

Useful sources of data

Biblographies. These are produced by the Geography Department, London School of Economics from 1964 onwards as Geographical abstracts. They are published in four sections: geomorphology; biogeography, climatology, pedology; economic geography; and social geography and cartography. They may be consulted in the libraries of universities, polytechnics and colleges of higher education for sources on the local area. For earlier papers on geomorphology the reference is Clayton, K. 1964, *A bibliography of British geomorphology* (London: Geo. Philip).

County Records Office. This will contain old maps such as those of towns, enclosures, tithes, estates and also large scale OS sheets for the county. It will contain the most extensive collection of publications on the county that is available and will include early histories, the County Report of Agriculture, and street directories such as Kelly's.

City or town Library. Here will be found the Census Report, copies of the local newspaper, rail and bus timetables and some of the statistics published annually by HMSO.

Parish statistics for agriculture may be obtained at a fee from the Ministry of Agriculture, Fisheries and Food, Government Buildings, Epsom Road, Guildford, Surrey. Information may also be available at the local divisional offices of the MAFF, or from the local secretary of the National Farmers Union.

Sources of maps. Local Ordnance Survey agents (see Yellow Pages of the telephone directory) will supply copies of 1:50 000, 1:25 000 and 1:10 000 sheets. David & Charles, Newton Abbot, Devon have reprinted the first edition OS one inch to one mile series. Details of local geological maps and memoirs in print may be obtained from the Institute of Geological Sciences, Exhibition Road, South Kensington, London.

Soil maps are available from The Soil Survey of England and Wales, Rothamsted Experimental Station, Harpenden, Herts, but only a few sheets have been published. The same problem of few published sheets applies also to the Second Land Utilization Survey map obtainable from Edward Stanford Ltd, 12–14 Long Acre, London, and to the Agricultural Land Classification maps from the Ministry of Agriculture, Fisheries and Food, Government Buildings, Tolcarne Drive, Pinner, Middlesex. Maps of very detailed surveys of central urban areas are available from Charles Goud Ltd, 18A Salisbury Square, Old Hatfield, Herts.

The local meteorological office will supply data on local temperature and rainfall, and may be consulted for other weather observations. For projects concerned with hydrology the *Surface water year book*, HMSO provides much data.

Several series of books are sufficiently numerous to be mentioned here as source material. These are: Hoskins, W.G. (ed.) 1956, *The making of the English landscape* (London: Hodder and Stoughton); Pevsner, N. (ed.) *Buildings of England* (London: Penguin); and the regional handbooks produced for the annual meetings of the British Association for the Advancement of Science.

Local Government Departments frequently give help to sixth-formers requesting information on local services. The Yellow Pages of the telephone directory should be consulted for their addresses.

Selected references

Books commonly used by sixth-formers which give basic subject matter:

Bradford, M.G. and W.A. Kent 1977. *Human geography.* Oxford: Oxford University Press.
Everson, J.A. and B.P. Fitzgerald 1969. *Settlement patterns.* London: Longmans.
Everson, J.A. and B.P. Fitzgerald 1972. *Inside the city.* London: Longmans.
Chisholm, M.D. 1962. *Rural settlement and land use.* London: Hutchinson.
Small, R.J. 1977. *The study of landforms,* 2nd edn. Cambridge: Cambridge University Press.
Sparks, B.W. 1972. *Geomorphology,* 2nd edn. London: Longmans.
Tidswell, V. 1976. *Pattern and process in human geography.* London: University Tutorial Press.
Weyman, D. and V. Weyman 1977. *Landscape processes.* London: George Allen & Unwin.

Books which have been used as sources of ideas for topics:
Cooke, R.U. and J.C. Doornkamp 1974. *Geomorphology in environmental management.* Oxford: Oxford University Press.
Geiger, R. 1966. *The climate near the ground,* 4th edn. Cambridge, Mass: Harvard University Press.
Gregory, K.J. and D.E. Walling 1974. *Drainage basin, form and process.* London: Edward Arnold.
Haggett, P. 1975. *Geography: a modern synthesis,* 2nd edn. New York: Harper & Row.
Kellman, M.C. 1975. *Plant geography.* London: Methuen.
And the journals:
Area. London: Institute of British Geographers.
Progress in human geography. London: Edward Arnold.
Progress in physical geography. London: Edward Arnold.

Collection of exercises containing more project topics:
Knapp, B.J. (ed.) 1980. *Practical foundations of physical geography.* London: George Allen & Unwin.

Books which explain techniques of geographical study:
Briggs, D.J. 1977. *Sources and methods in geography: Soils. Sediments.* London: Butterworth.
Hanwell, J.D. and M.D. Newson 1973. *Techniques in physical geography.* London: Macmillan.
Morgan, M.A. 1979. *Historical sources in geography.* London: Butterworth.
Toyne, P. and P.T. Newby 1971. *Techniques in human geography.* London: Macmillan.

A reliable statistics textbook:
Hammond, R. and P.S. McCullagh 1978. *Quantitative techniques in geography, 2nd edn.* Oxford: Oxford University Press.

For advice on questionnaires:
Oppenheim, A.N. 1968 *Questionnaire design and attitude measurement.* London: Heineman.

For help with diagrams:
Monkhouse, F.J. and H.R. Wilkinson 1962. *Maps and diagrams,* 2nd edn. London: Methuen.

A more exhaustive list of references is to be found in Bates, J.N. *Further developments in geography teaching* – a 1979 bibliography, which is available from the Geographical Association, 343 Fulwood Road, Sheffield S10 3BP.

Index